SMALL ENGINES
AND OUTDOOR POWER EQUIPMENT

A CARE & REPAIR GUIDE FOR: **LAWN MOWERS, SNOWBLOWERS & SMALL GAS-POWERED IMPLEMENTS**

Edited by Peter Hunn

COOL SPRINGS PRESS
Home and Garden Experts™
MINNEAPOLIS, MINNESOTA

First published in 2014 by Cool Springs Press, a member of the
Quarto Publishing Group USA Inc., 400 First Avenue North, Suite
400, Minneapolis, MN 55401

Cool Springs Press titles are also available at discounts in bulk
quantity for industrial or sales-promotional use. For details write
to Special Sales Manager at Cool Springs Press, 400 First Avenue
North, Suite 400, Minneapolis, MN 55401 USA. To find out more
about our books, visit us online at www.coolspringspress.com.

Library of Congress Cataloging-in-Publication Data

Small engines and outdoor power equipment : a care & repair
guide for lawn mowers, snowblowers & small gas-powered
implements / editors of Cool Springs Press.
    pages cm
 ISBN 978-1-59186-587-2 (pbk)
 1. Small gasoline engines--Maintenance and repair. 2.
Gardening--Equipment and supplies--Maintenance and repair. I.
Cool Springs Press. II. Title: Small engines and outdoor power
equipment.

 TJ790.S5827 2013
 621.43'40288--dc23

                        2013028515

Acquisitions Editor: Mark Johanson
Design Manager: Brad Springer
Layout: Laurie Young
Cover Design: Simon Larkin

Printed in China
10 9 8 7 6 5 4 3 2

Contents

Introduction

Unless you're among the small but enthusiastic group of "motorheads" who collect, troubleshoot, and restore vintage lawn mowers, chain saws, garden tractors, snowblowers, or old outboard motors as a hobby, you probably opened this book because some small engine in your life just isn't running right. Maybe it's hard to start or sometimes doesn't start at all. Perhaps your small engine runs, but sounds strange; like it's got real problems somewhere deep down inside.

Quite possibly you're wishing you had the mechanical equivalent of a "green thumb" so you could save money on small-engine repairs. With most power equipment shops charging hourly repair rates comparable to those of a luxury car dealership's service department, it is no wonder why many folks who don't consider themselves particularly handy get so darn frustrated by their persnickety weed-whacker or messed-up mower that they feel there's no alternative but to buy a new one. With all due respect to those who lack confidence in their ability to keep their little motors purring proudly season after season, *Small Engines & Outdoor Power Equipment* is written to prove them wrong.

ORGANIZATION OF THE BOOK

This book walks you through lots of "real world" mechanical cures and problem prevention, based on the most widely used small-engine designs. You can find the step-by-step directions for your specific maintenance or fix-up project in the Table of Contents (page 3) or by delving right into one of the sections listed below. Each offers safety tips, key to efficient diagnosis, and proficient "wrenching."

"Things That Make A Small Engine Run" (pages 6 to 33) identifies parts of the small engine and outlines how they are arranged into systems: They are the **compression** system, **fuel** system, **ignition** system, **lubrication and cooling** system, **braking** system, and the **electrical** system. This chapter also identifies good tools to have on hand for repair work, as well as some pointers on safe small-engine servicing.

"Motor Medical School 101" (focuses on troubleshooting that poor old motor's funny noises   or maybe getting it to say something nice to you again (pages 34 to 41).

"Easy But Important Maintenance" (pages 42 to 65) offers a convenient schedule and checklist for inspecting and changing the oil, spark plugs, and filters.

"Doctoring The Ailing Engine" **"Basic Engine Repairs"** (pages 66 to 95) and **"Advanced Repairs"** (pages 96 to 137), guide you through being able to and repair of the most common problems. Also covered are engine maladies, that some might consider to be a bit more challenging.

"2-Stroke Power Equipment" (pages 138 to 143) introduces you to the basics of 2-stroke engine operation and then guides you through basic maintenance and repair common to outdoor power equipment implements such as chain saws, string trimmers, snowblowers, and gas blowers.

Things That Make a Small Engine Run

Defining the Small Engine

The biggest difference between small gasoline engines and other types of fuel-burning motors is their small capacity and simplicity of design. Small engines generate modest amounts of power—generally 2 to 25 horsepower—compared to a typical family car boasting 150 or more horses. And a small engine's compact size also makes it easier to maintain and repair.

A typical 4-stroke, overhead valve gas engine. At 6.5 horsepower it is of average power for a lawn mower.

Because small engines are designed for relatively simple tasks like cutting grass and blowing snow, their construction is uncomplicated compared to other engines. Unlike cars and other vehicles that frequently accelerate, slow down, or idle for long periods, small engines are usually run at constant speed or change speed slightly when encountering modest changes in the "load," such as when a lawn mower hits a patch of thick grass, or when a snowblower tries to digest a pile of packed snow recently shoveled off the roof and into the driveway.

Also, unlike car engines, small engines don't have to be wedged between a radiator and firewall in an automotive chassis or get linked to countless computers and other electrical devices. This makes small engine parts easier to install, adjust and remove. In many cases, you can reach even the most hidden small engine parts with just a few turns of a wrench.

ESSENTIALS OF THE 4-STROKE ENGINE

Except for most chain saws, string trimmers, leaf blowers, and light-duty snowblowers that use 2-stroke (also known as 2-cycle) engines (see *2-Stroke Power Equipment* on page 138), the majority of small power plants are of the 4-stroke (or 4-cycle) variety. These diminutive 4-strokes follow the same essential operating principles as automotive engines. To use a bakery metaphor, the little mills on lawn mowers would be considered cupcakes compared to the far-larger, multitiered confection created for a fancy wedding reception. No matter the intended application, however, here, in its simplest form, is how a 4-stroke engine works:

When you pull the rope, known as a rewind cord, or use your electric starter, precise amounts of fuel and filtered air mix together in the carburetor. The mixture rushes into the engine to be compressed, ignited, and burned in a controlled process known as internal combustion. This produces hot gases. As these gases expand, they push a smooth, well-lubricated cylindrical component, known as the piston. The piston, in turn, drives the crankshaft, the arm that spins a blade or performs other work. Moving valves on a 4-stroke engine (and stationary ports in the cylinder wall on a 2-stroke motor) let air and fuel into the combustion chamber above the piston and allow spent gases to exit through the muffler. Pulling vigorously on that cord or turning the starter key gets this process moving. When it works correctly, it coaxes the fuel-air atomization, fuel-air compression, ignition/spark, and exhaust of burned gases to work on its own. It is designed to become self-sustaining from the time the engine starts until the moment it stops. Timed electrical surges cause the spark plug to fire repeatedly inside the combustion chamber, igniting each fresh supply of air and fuel and producing gases that continually drive the piston and crank shaft. All the while, oil from the crankcase and air circulate to keep engine temperatures within an acceptable range, and a governor monitors changes in the workload and adjusts engine speed accordingly. There's a lot being asked of a small engine when the starter is suddenly engaged. That's why an engine might sound like it's taking a second or two to fully come to life, the way somebody might when roused out of a warm, comfortable bed when the clock radio alarm fires up on a winter's morning. Two-cycle engines complete this "wake-up" routine in two piston "strokes" (one up, one down, or one 360-degree rotation of the crankshaft). For a 4-cycle engine, the whole process involves four motions of piston travel (two up strokes and two down strokes, or two full crankshaft rotations.

Five basic small-engine systems

There are five systems at work in every small engine: fuel supply, compression, ignition, lubrication and cooling, and governor (speed control). Each of these systems is explained in depth in the following pages.

Two other common systems—starter systems, which require a battery, starter motor, and electrical recharging system, and brake systems, which stop the engine if you let go of the controls—are also discussed in detail. In short, these five systems generate the power to spin a blade, turn a wheel, or perform other work, while the two others—starters and braking systems—increase safety and convenience. The following pages will familiarize you with the major parts in these systems and the essentials of how they work.

Parts of the Fuel System

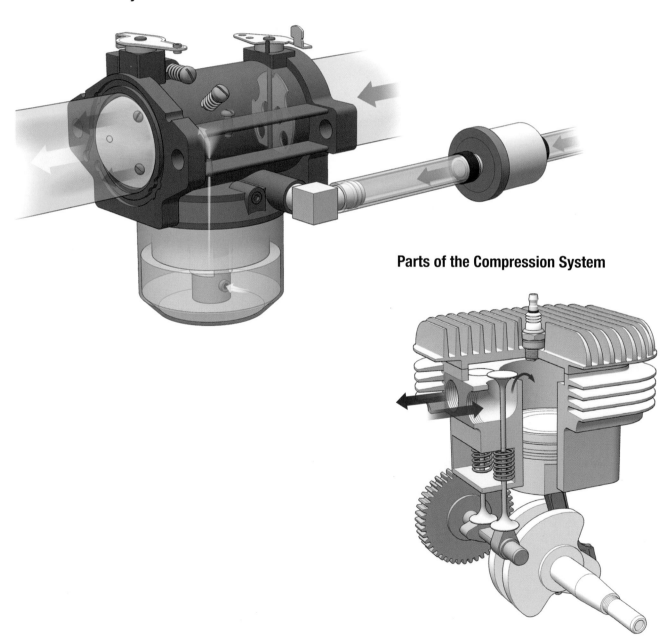

Parts of the Compression System

Parts of the Ignition System

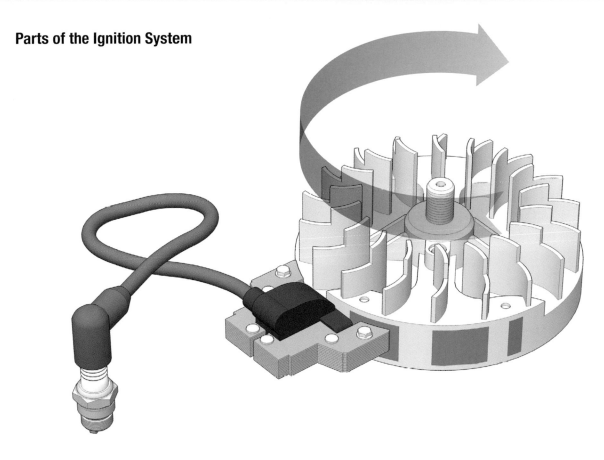

Parts of the Lubrication System

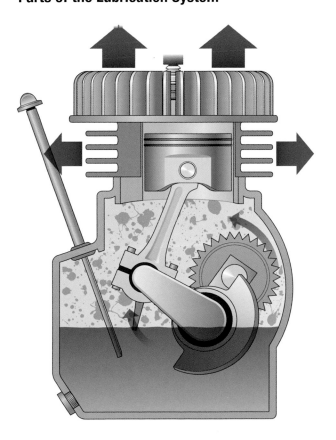

Parts of the Cooling System

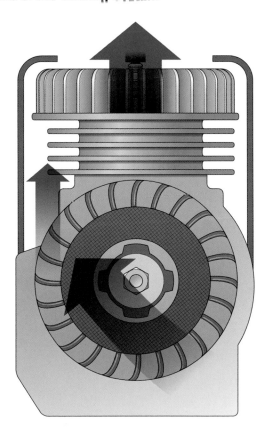

ENGINE COMPONENTS AND THEIR FUNCTION

Here's how the components in your engine interact to start and maintain the combustion process:

- The rewind cord is pulled to start the combustion process. On some models, a starter motor replaces the rewind, drawing on battery power to start the engine.

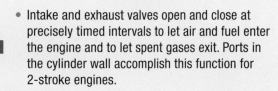

- Revolving magnets work in conjunction with the ignition armature and spark plug to produce a spark in the combustion chamber.

- The carburetor draws in fuel from the fuel tank and outside air to form a combustible vapor that is fed into the combustion chamber.

- Intake and exhaust valves open and close at precisely timed intervals to let air and fuel enter the engine and to let spent gases exit. Ports in the cylinder wall accomplish this function for 2-stroke engines.

- The piston is pushed through the cylinder by the force of expanding gases. The piston's motion causes the crankshaft to turn. Momentum then carries the piston back toward the top of the cylinder.

- In 4-strokes (4-cycle), oil stored in the crankcase circulates through the engine to lubricate key components like the piston and crankshaft and to provide generalized cooling by drawing away heat from internal engine surfaces.

- Two-stroke engines receive lubrication via oil that the user mixes into the gasoline. While some might see religiously following 2-cycle engine manufacturers' specific mix ratios for each fill-up as a nuisance, fans of this type of power-plant correctly argue that the properly fueled 2-stroke is essentially benefiting from a fresh oil change with each use—a luxury unheard-of in 4-stroke circles.

- A flywheel brake and stop switch are included on engines for equipment such as mowers that require constant supervision. The two components are designed to stop the engine if you release the controls.

- An air vane or flyweights monitor engine rpms so the governor can maintain the selected engine speed.

- Cooling fins help reduce engine temperatures when air circulates across the hottest engine surfaces.

TYPICAL SMALL ENGINES FOR CONSUMER POWER EQUIPMENT

7 HP Horizontal Shaft Single Cylinder

Uses: Small outdoor power equipment (edgers, pressure washers, shredders, tillers, etc.)

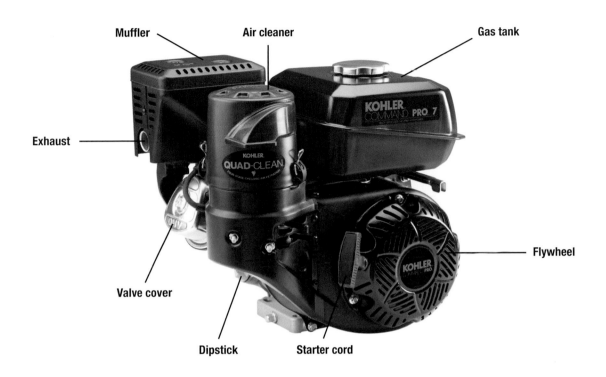

Muffler — Air cleaner — Gas tank — Exhaust — Flywheel — Valve cover — Dipstick — Starter cord

26 HP Vertical Shaft V-Twin

Uses: Lawn tractors

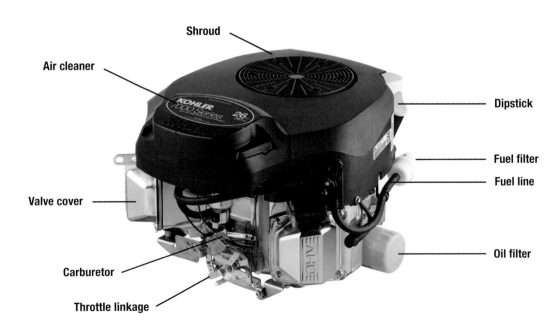

Shroud — Air cleaner — Dipstick — Fuel filter — Fuel line — Valve cover — Oil filter — Carburetor — Throttle linkage

The Compression System

The inventors of the first internal combustion engines discovered that fuel burns more efficiently if compressed in a sealed chamber before burning it. Compression of the air-fuel mixture in the small 4-stroke engine begins as the intake valve closes (or as the piston covers the cylinder ports on a 2-stroke mill). The trapped vapors are pushed toward the cylinder head by the piston and compressed into a space about one-sixth their original volume. The exact amount of compression is an indicator of an engine's efficiency. That's why a tightly sealed combustion chamber is so important for good engine performance.

VALVES ON A 4-STROKE ENGINE

Valves located in the combustion chamber let fuel vapors and air enter the cylinder and let exhaust gases exit at precisely timed intervals. A typical 4-stroke small engine contains one intake valve and one exhaust valve per cylinder, and most small 4-stroke engines have one cylinder and use an L-head (or flathead) design, where the valves are installed in a valve chamber next to the piston. Overhead valve (OHV) designs offer greater efficiency, however, and are increasingly popular with consumers. In this design, the valves are located in the cylinder head directly in line with the piston and are moved by pivoting rocker arms.

PISTON

The piston rides through the cylinder, much as a plunger rides through the chamber in a hand-operated air pump. At the appropriate moment, the cylinder is sealed so that the air-fuel mixture is compressed as the piston moves toward the cylinder head. When the mixture is ignited, rapidly expanding gases force the piston back down through the cylinder. While 4-stroke engines typically have flat-topped pistons, 2-stroke owners getting into their engines will discover pistons with an exhaust deflector cast into the piston crown.

RINGS

The piston diameter is narrow enough to permit a thin space around it for a coating of oil. Flexible piston rings, installed in grooves in the piston, work in concert with the oil to create a seal between the piston and the cylinder wall, thus ensuring good compression. As the piston is pushed down through the cylinder by expanding gases, a connecting rod transfers the force of those gases to the flywheel. It's the flywheel's momentum that perpetuates the engine's 4-stroke cycle.

COMPRESSION PROBLEMS

Too little or too much compression can damage pistons, rings, valves, valve guides, valve seats, and the cylinder wall. Loss of recommended compression (measured in pounds per square inch) can prevent an engine from starting at all. On a 4-stroke, if an exhaust valve leaks, exhaust can back up into the cylinder, causing premature wear. Too much compression can cause the air-fuel mixture to burn too fast, causing knocking or pinging. Excess compression can also leave carbon deposits that further aggravate problems.

PARTS OF THE COMPRESSION SYSTEM

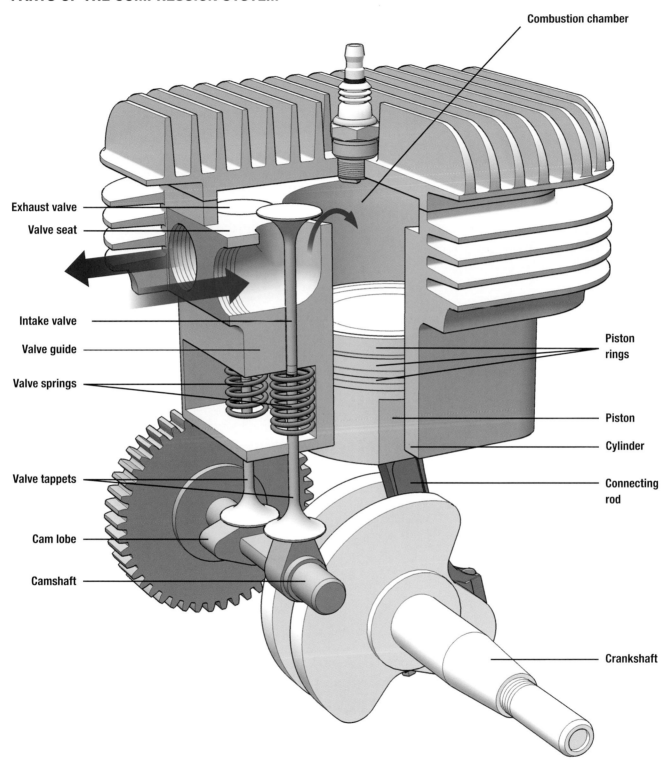

Combustion chamber

Exhaust valve

Valve seat

Intake valve

Valve guide

Valve springs

Valve tappets

Cam lobe

Camshaft

Piston rings

Piston

Cylinder

Connecting rod

Crankshaft

The 4-Stroke Cycle

In most engines, including the one you'll find on your car, and most probably your lawn mower, tractor, tiller, wood chipper, or other outdoor power equipment, combustion occurs in a 4-stroke cycle.

The 4-stroke cycle involves four distinct piston strokes (intake, compression, power, and exhaust) that occur in succession. For each complete cycle, there are two complete rotations of the crankshaft.

THE INTAKE STROKE

During the intake stroke, a mixture of air and fuel is introduced to the combustion chamber. The intake valve is open and the piston moves from Top Dead Center (TDC) to Bottom Dead Center (BDC).

To understand what happens next, think of the suction produced like a syringe drawing liquid. This happens because as the plunger inside slides toward the handle, it creates a low-pressure area at the tip. A piston performs the identical task. As the piston moves toward BDC, it creates a low-pressure area in the cylinder and draws the air-fuel mixture through the intake valve. The mixture continues to flow, due to inertia, as the piston moves beyond BDC. Once the piston moves a few degrees beyond BDC, the intake valve closes, sealing the air-fuel mixture inside the cylinder.

THE COMPRESSION STROKE

Compression occurs as the piston travels toward TDC, squeezing the air-fuel mixture to a smaller volume. The air-fuel mixture is compressed for a more efficient burn and to allow more energy to be released faster when the mixture is ignited. Think about the warning label on pressurized spray cans: Keep contents away from fire. This is not only because the contents are flammable, but because pressurization makes them potentially explosive.

If an engine has to perform so much work just to bring the air-fuel mixture to the point of combustion, where does it find the ability to perform work? This ability derives from the fact that the energy required for compression—and stored in the flywheel—is still far less than the force produced during combustion. In a typical small engine, compression requires one-fourth the energy produced during combustion. The surplus drives the power stroke.

THE POWER STROKE

The engine's intake and exhaust valves are now closed. At approximately 20 degrees before TDC, the spark plug initiates combustion, creating a flame that burns the compressed air-fuel mixture. The hot gases produced by combustion have no way to escape, so they push the piston away from the cylinder head. That motion is transferred through the connecting rod to apply torque to the crankshaft.

THE EXHAUST STROKE

As the piston reaches BDC during the power stroke, the power stroke is completed. The exhaust valve opens, allowing the piston to evacuate exhaust as it moves, once again, toward TDC. With the chamber cleared of exhaust, the piston reaches TDC. An entire cycle is complete.

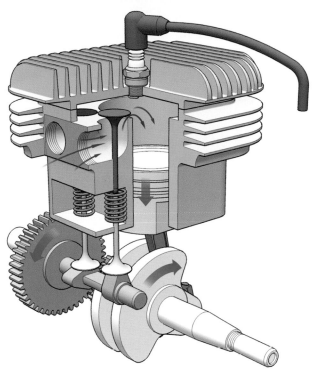

1. Intake Stroke

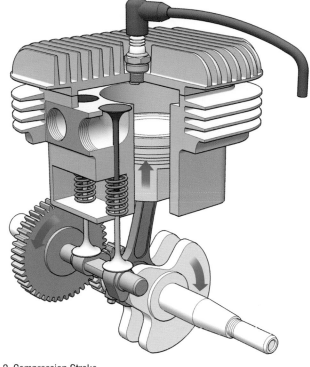

2. Compression Stroke

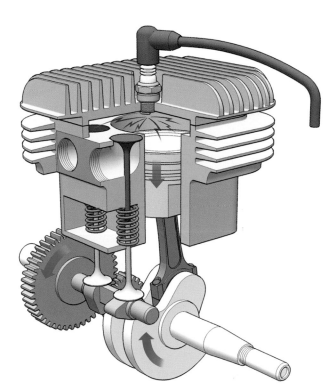

3. Power Stroke

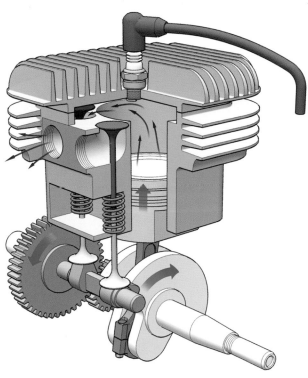

4. Exhaust Stroke

Overhead Valves (OHV): The New Standard

Locating the valves next to the piston is just one way to configure an engine. Engineers figured out long ago that they could gain a significant advantage in many higher-horsepower engines by installing the valves in the cylinder head so that they face the piston. Pivoting rocker arms—moved by push rods—open the valves.

One of the main advantages of overhead valve design is a more symmetrical combustion chamber, resulting in a more efficient burning of the air-fuel mixture. You may see the letters "OHV" imprinted on the shroud of your engine to indicate the use of overhead valves.

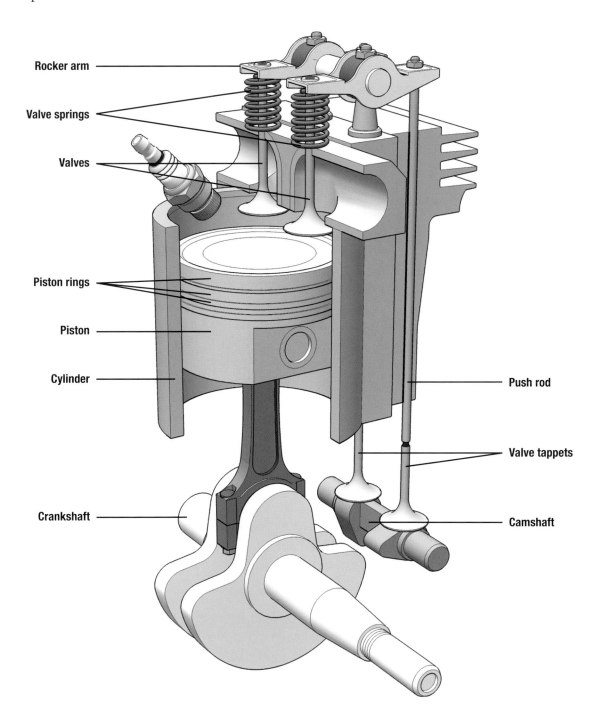

- Rocker arm
- Valve springs
- Valves
- Piston rings
- Piston
- Cylinder
- Crankshaft
- Push rod
- Valve tappets
- Camshaft

Fuel System

The most common repairs involve the fuel system, which includes the fuel tank, fuel filter, fuel line, fuel pump (on some models), and carburetor.

You may have heard people jokingly comment that the fuel in the tank is so low the engine is running on fumes or "vapors." Technically, they're right. Gasoline won't burn in its liquid state; it must be converted to a vapor first. The vapors that burn in your small engine are formed from a mixture of fuel (typically gasoline) and air. And you need the right amount of fuel and the right amount of air to maintain whatever engine speed you select. The best way to understand the fuel system is to begin at the tank.

FROM THE FUEL TANK TO THE CARBURETOR

Locate the fuel tank on your engine. If you have an older engine, its tank is probably made of steel or aluminum (subject to rust or corrosion). Newer tanks are made of plastic and are built into the molded plastic shroud over the engine. Now, look for a fuel line, a hose connected to one side of the tank. The fuel line carries the fuel to the carburetor, a mixing chamber that contains a throttle and (if equipped) a choke, attached to the equipment controls.

On most engines, the force of gravity carries the fuel through the fuel line. However, if the fuel tank is mounted low on the engine, gravity may not do the trick. In this case, a fuel pump uses low pressure in the crankcase to pump fuel. The pump is located between the tank and the carburetor or in the carburetor itself. Some engines eliminate the need for either a fuel line or pump by mounting the carburetor directly on the fuel tank and using a pick-up tube in the tank to draw fuel.

Locating the fuel tank is not difficult at all. Modern push mowers usually have plastic tanks that are built into the shroud. On smaller handheld power equipment the fuel tank is often opaque plastic and mounted at the bottom of the tool.

Into the Carburetor

On most engines, fuel from the fuel line enters the carburetor's fuel bowl, a reservoir where a float (similar to the float ball in a toilet tank) regulates the fuel level. From there, a metering device called a jet lets fuel into the emulsion tube inside the pedestal, where fuel and air first mix. (Older models include an adjustable jet; newer models contain a fixed jet.) Fuel travels through the emulsion tube to the main passageway in the carburetor, called the throat or venturi, where further mixing occurs. If your carburetor is a tank-mounted type, fuel from the tank may be supplied directly to the emulsion tube, without the need for a float.

THE ROLE OF THE THROTTLE

At one end of the throat is a throttle plate. The throttle plate is connected to your equipment control lever (often referred to as the throttle) and opens or closes to increase or decrease engine speed. As the throttle plate opens, more air is drawn into the carburetor. Air flow, in turn, determines how much fuel is delivered for combustion. Many carburetors have an idle speed screw to stop the throttle from closing too far at low speed, and an idle mixture screw, which increases or decreases air and fuel flow to prevent a stall.

USING THE CHOKE

A throttle works fine in warm weather. But when it's cold, fluids don't vaporize as easily. The engine may need extra fuel to start. This is the role of the choke plate or primer. They compensate for the cold by increasing the fuel-to-air ratio. The choke is located in the throat between the air filter and the throttle plate. Closing the choke reduces air flow. Low pressure created inside the engine keeps the fuel flowing. The use of the choke "enriches the mixture." It's not an effective way to run an engine all the time, but it helps a cold engine start. Once the engine reaches its normal operating temperature range, you can open the choke to let in more air, for a cleaner, more efficient burn.

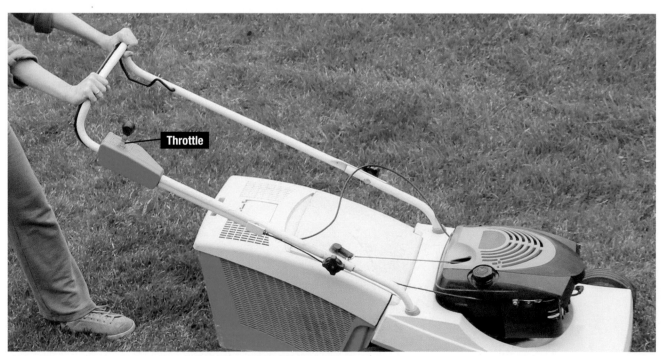

Throttle

Engine speed is controlled by the throttle, a grippable lever that opens and closes the throttle plate.

PARTS OF THE FUEL SYSTEM

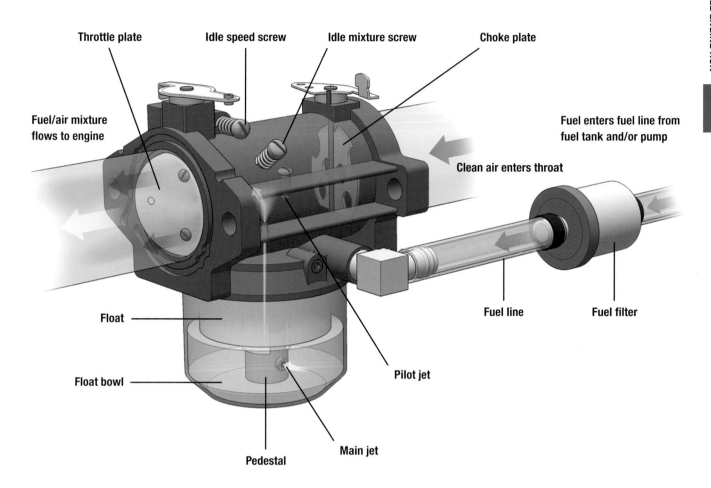

Throttle plate   Idle speed screw   Idle mixture screw   Choke plate

Fuel/air mixture
flows to engine

Fuel enters fuel line from
fuel tank and/or pump

Clean air enters throat

Float

Fuel line   Fuel filter

Float bowl

Pilot jet

Pedestal   Main jet

UNDERSTANDING CARBURETION

In essence, a carburetor is a passageway that draws in air and fuel and supplies a mixture of the two to the cylinder. It's the narrowing of the passageway—called the throat or venturi—that causes the carburetor to draw in the two components the engine needs for combustion. Basic physics tells us that air speed will increase at the narrow point and air pressure will drop. Since fluids flow to low-pressure areas, fuel from the bowl or tank is drawn into the throat, mixing with air to form a combustible vapor.

Ignition System

Ignition system parts include the flywheel, ignition armature, magnets, spark plug, and spark plug lead.

The ignition system plays a central role in the starting of your small engine. Whether you start the engine with a tug on the rewind rope or the turn of a key on an electric starter motor, you're relying on the ignition system to produce a spark inside the combustion chamber.

The ignition system includes magnets mounted in the surface of the flywheel, and an ignition armature mounted adjacent to the flywheel, containing copper wire windings. It also includes the spark plug lead (attached to the armature) and the spark plug.

When you pull on the rewind rope, you are turning the flywheel, a heavy metal wheel located under the blower housing. With each turn, the magnets mounted in the surface of the flywheel pass the ignition armature, inducing electrical flow that produces a high-voltage spark at the tip of the spark plug. The ignition system is coordinated with the timing of the piston and the motion of the valves so that the spark will ignite the air-fuel mixture in the combustion chamber just as the piston reaches the point of maximum compression in each engine cycle.

Once the engine is running, the flywheel's inertia keeps the crankshaft spinning until the piston's next power stroke, while the flywheel magnets induce voltage in the armature to keep the spark plug firing.

SOLID-STATE IGNITION SYSTEMS

It takes 10,000 to 20,000 volts of current to produce a spark at the tip of a spark plug. That's enough to give a person a powerful jolt. Today's ignition systems accomplish this using a tiny transistor in the ignition armature. Each time the magnets approach, the transistor establishes an electrical circuit, also called "closing" a circuit. The 2 to 3 amps of current produced are then converted to high-voltage current that travels through the spark plug lead to the spark plug.

BREAKER POINT IGNITION SYSTEMS

Breaker point systems are found on small engines built until the early 1980s. They function much like solid-state ignitions, but use a mechanical switch, instead of a transistor, to close the electrical circuit required to produce a high-voltage spark at the spark plug tip. A pair of nickel-plated breaker points remain apart for most of the 4-stroke cycle (and half that time on a 2-stroke engine). A flat spot machined into the crankshaft causes one of the points to pivot temporarily, closing the gap between the two and closing a circuit.

BUYING SYSTEM PARTS

Ignition systems are designed to operate efficiently with specifically designed components and should not be altered. When you need to replace parts, use the original manufacturer's replacement parts or quality "generics" made with your particular engine's specifications in mind.

Parts of the Ignition System

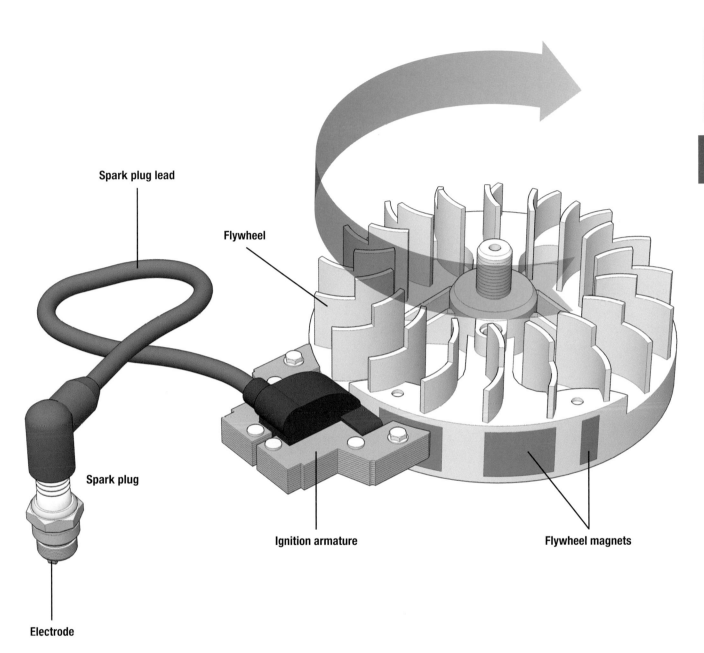

Spark plug lead

Flywheel

Spark plug

Ignition armature

Flywheel magnets

Electrode

Lubrication & Cooling System

Lubrication parts of a 4-stroke mill include the crankcase, drain plug, oil fill cap, oil dipper or slinger, and oil dipstick. Air cooling parts include the cylinder head cooling fins and flywheel fins.

Exhaust gases and radiant heat emitted from engine components carry off much of the heat produced by a small engine, but not enough to keep an engine running reliably. The lubrication and cooling system is designed to handle that task. As a lubricant, oil not only carries away heat, it reduces a major source of heat—friction between engine components. Airflow also serves a secondary function on some engines, triggering the air vane in a pneumatic governor.

REDUCING FRICTION WITH OIL

Viscosity is the most important quality of engine oil. It is a measure of an oil's ability to resist motion. This quality is critical to oil's performance, since moving parts constantly try to push oil, much as a plow pushes snow. Oil must resist this force so it can maintain a continuous film that keeps the parts themselves from touching.

While viscosity allows oil to cling to surfaces and resist the snowplow effect, it also reduces the ability to flow at low temperatures or within tight clearances. A more viscous oil also takes longer to reach its optimal temperature. And while oil grades are generally a compromise that tries to anticipate typical operating conditions, a common recommendation for 4-stroke small-engine oil is SAE (Society of Automotive Engineers) 30 four-cycle oil.

Some oils are altered to make them less viscous during the winter. These multi-viscosity oils have ratings such as SAE 10W-30. The 10W indicates a lower winter viscosity. At normal operating temperatures, the oil acts like SAE 30 oil. Two-stroke oil is typically divided into categories for air-cooled 2-cycle engines or for water-cooled motors such as outboards. The former runs hotter than the latter, but synthetic oils are arguably best for both. No matter the engine type, always follow the recommendation of your engine's owner's manual when selecting the proper oil.

Engine oil is formulated differently for different types of outdoor power equipment, but this SAE 30 oil will work in all 4-cycle engines, lawn mowers or otherwise.

USING MULTI-VISCOSITY OIL

Multi-viscosity oil, such as SAE 10W-30, is designed to work well in cold weather. It's not the best choice if you typically operate your equipment at or above 40°F. In warm weather, multi-viscosity oil is likely to cause premature carbon build-up and loss of engine power. Two-stroke engines require oil specifically labeled for 2-cycle engines.

GETTING OIL TO CIRCULATE

Most 4-stroke small engines rely on the splashing motion of a dipper or slinger in the crankcase to distribute oil. On a horizontal crankshaft engine, a dipper is attached to the connecting rod. It picks up oil in the oil reservoir located in the crankcase and spreads it across bearing surfaces as the piston travels through the cylinder.

A slinger is used on many vertical crankshaft engines. It consists of a spinning gear with paddles cast into the plastic gear body. Part of the slinger is submerged in the oil. As the crankshaft turns, the slinger disperses oil throughout the crankcase.

KEEPING A 4-STROKE'S OIL CLEAN

Small engines designed for tractors and other heavy-duty equipment may include an oil filter. The pleated paper inside removes dirt, metal particles and other foreign matter that accumulates in the oil. If the paper becomes clogged, oil is rerouted through a spring-loaded bypass valve to ensure lubrication even when oil is very dirty. Even if your engine has an oil filter, you need to inspect the oil every time the engine is run to make sure the level is correct and the oil still has its clean amber color.

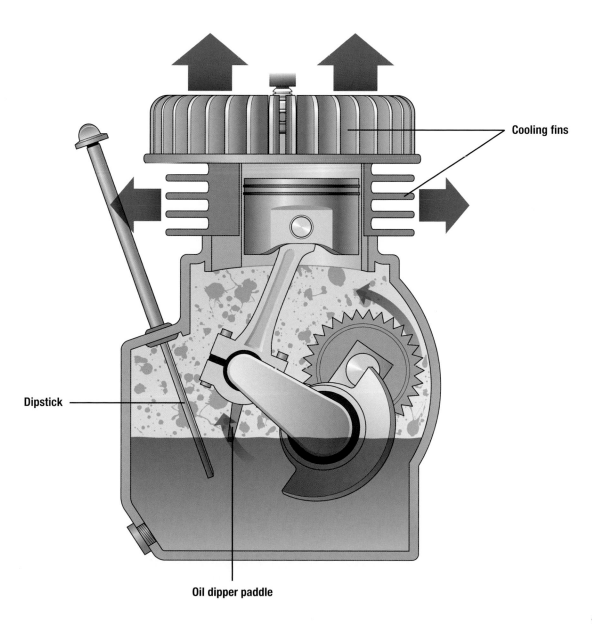

Cooling fins

Dipstick

Oil dipper paddle

COOLING WITH AIR

An engine relies on air circulating around engine parts to maintain an acceptable engine temperature. Fins on the outside of the cylinder block and cylinder head improve the engine's cooling ability the way pipes do on a car radiator—by increasing the surface area that radiates heat and is exposed to cool air. Although air is not the most efficient way to transfer heat, it is plentiful and usually offers a substantial cooling effect.

A different set of fins—the flywheel fins—are also an important cooling feature; as they spin they distribute air to many engine parts. The blower housing and air guides route air to the flywheel fins.

On some models, a rotating screen over the flywheel prevents grass and other debris from clogging the flywheel fins. The screen blocks debris from entering or cuts it into smaller, less harmful particles.

ADDITIONAL COOLING COMPONENTS

Some engines require additional air cooling and contain a cooling air plenum, a duct that provides a separate means for outside air to enter the engine. Some contain an air discharge that directs hot air away from the engine.

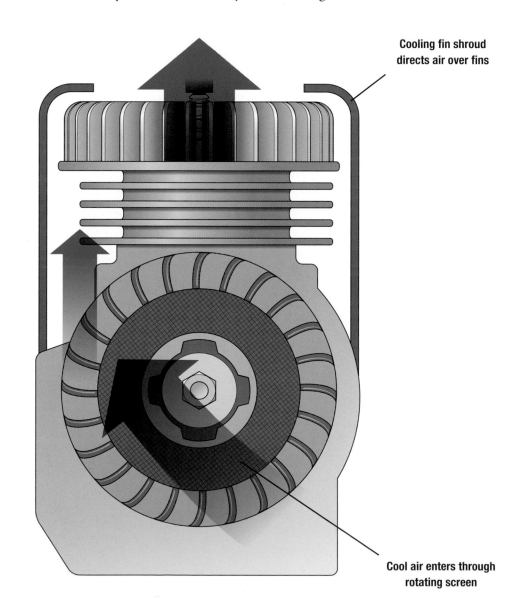

Cooling fin shroud directs air over fins

Air flows from flywheel to cooling fins

Cool air enters through rotating screen

GOVERNOR SYSTEM

A mechanical governor includes gears, a governor crank, governor springs, flyweights, and linkages.

The governor system is like a cruise control system. It keeps the engine running at the speed you select, regardless of changes in the load. You can think of the load as the amount of work the engine must perform: for a mower, the height of the grass; for a tiller, the depth of the tines; for a chipper, the thickness of the branches. Without a governor, you would need to adjust the throttle manually each time your lawn mower ran across a dense patch of grass. A governor does the job for you by detecting changes in the load and adjusting the throttle to compensate.

The governor system behaves like an unending tug of war between one or two governor springs, which pull the throttle toward the open position, and a spinning crankshaft, which tries to close the throttle. When the load on the engine increases—a typical example is when you move your running lawn mower from the driveway to the grass—crankshaft revolutions drop. But the governor spring is still tugging, causing the throttle plate to open. In response, a larger volume of air-fuel mixture enters the carburetor, increasing engine speed to compensate for the increased load. The crankshaft speeds up, and the tug of war resumes, until a new equilibrium is achieved. With each change in load, the tension between the governor spring and the load brings about a new equilibrium, known as the engine's governed speed. Neither side wins until the engine is shut off. At that point, without the crankshaft spinning, the governor spring pulls the throttle to the wide-open position.

Two types of governor are common on small engines—mechanical and pneumatic.

Mechanical governor

A mechanical governor uses gears and flyweights inside the crankcase as a speed-sensing device that detects changes in the load and adjusts the throttle accordingly.

Pneumatic governor

A pneumatic governor uses a movable air vane, made of metal or plastic, as a speed-sensing device by registering the change in air pressure around the spinning flywheel.

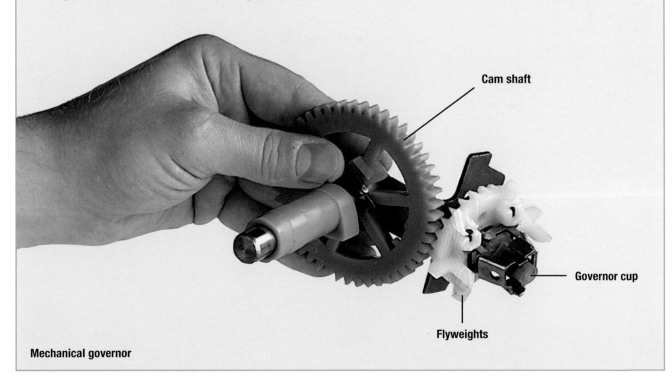

Cam shaft

Governor cup

Flyweights

Mechanical governor

Braking System

Small-engine braking systems include a brake band or pad, brake bracket, brake cover, brake spring, stop switch, and stop switch wire, all designed to stop the engine quickly.

In the old days, you could leave your lawn mower or tractor idling while you stepped into the garage for a rake. Today's small engine contains an automatic shut-off system designed to protect you and others in the area by stopping the engine any time you let go of the controls or climb off of the equipment.

Many small engines contain a brake that applies pressure to the smooth surface on the flywheel. The brake's surface area varies in size, depending on the equipment. Some models use a brake pad. Others use a brake band, applied to a larger area on the flywheel's surface. Both are highly effective when properly maintained.

Most engines contain one or more stop switches wired between the engine's ignition system and engine and equipment components. You can trigger the switch by releasing the brake bail or removing the grass discharge unit on a lawn mower or by standing up from the seat of a lawn tractor, triggering a switch under the seat.

The switch cuts power to the engine by grounding one of the copper windings in the ignition armature. When the brake bail is released, a wire attached to the armature is grounded against a metal engine part, stopping the engine. If the engine is equipped with a brake, the brake pad or band simultaneously applies pressure to the flywheel. When properly maintained, the two components of the braking system stop the engine within three seconds.

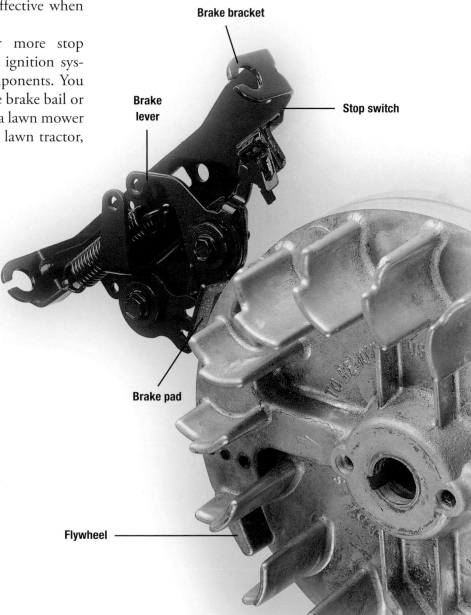

Brake bracket

Brake lever

Stop switch

Brake pad

Flywheel

Electrical System

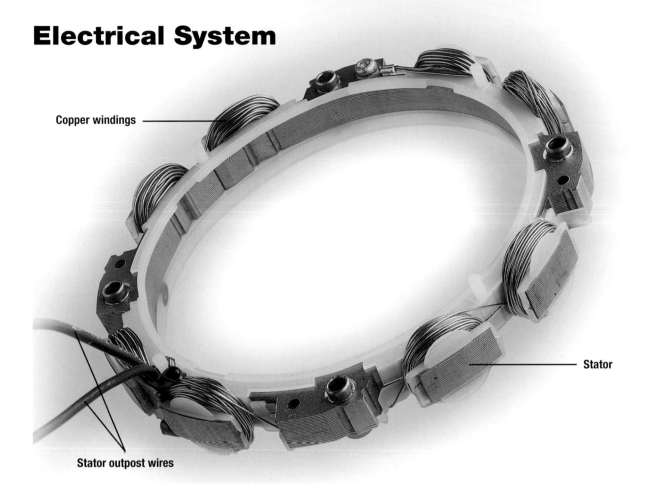

Copper windings

Stator

Stator outpost wires

A small-engine electrical system typically consists of an alternator, rectifier, regulator, and 12-volt battery. The alternator itself consists of an assembly of one or more copper windings—collectively known as the stator—and a set of magnets. Like the ignition system, the alternator creates a moving magnetic field to induce current. Most stators consist of a band of non-adjustable windings mounted under the flywheel and a set of magnets cemented to the inside surface of the flywheel. On some engines, the stator consists of an adjustable armature mounted outside the flywheel that relies on the same magnets as the ignition armature to charge the battery. The result is longer periods of time between surges of voltage and current. Limited amounts of DC voltage and current are produced, and a capacitor is used to handle fluctuations in the voltage output.

ALTERNATING CURRENT VS. DIRECT CURRENT

An electrical system can be set up to produce either alternating current (AC) or direct current (DC). If your equipment runs lights and no battery or other electrical devices, the alternator operates much like the generator on a bicycle wheel, keeping the lights running with AC as long as the bike wheel (or the engine crankshaft) is spinning. If your equipment includes a battery and various electrical devices, a rectifier is attached to the alternator to convert AC power to DC so it can be stored in the battery. DC power can run lights even when the engine is off, as well as a starter motor, electric cutting blade clutch, winch, and other devices. Some engines supply AC for lights and DC for other devices.

Engines that operate at high speeds also require a regulator or a combined regulator/rectifier to maintain a steady voltage output.

Tools to Keep Your Small Engine Happy

A combination of general purpose and specialty tools will fill your mechanic's toolbox

The right tools make your maintenance and repair projects easier and more successful. Have these tools available:

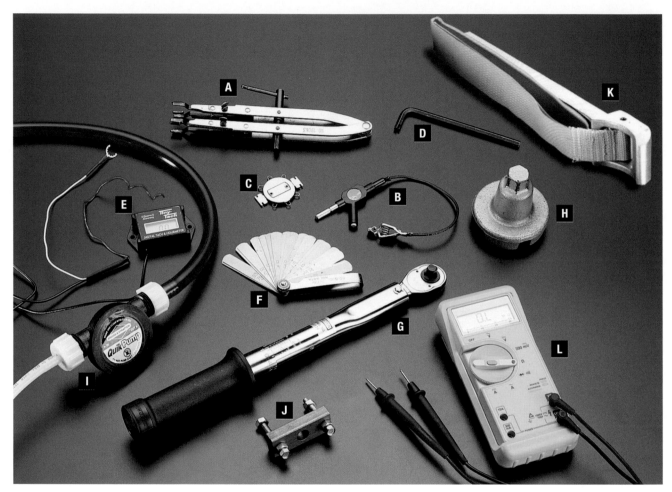

Small-engine tools

(A) Valve spring compressor
(B) Spark tester
(C) Spark plug gauge
(D) Tang bending tool

(E) Tachometer
(F) Feeler gauges
(G) Ratchet torque wrench
(H) Starter clutch wrench

(I) Oil evacuator pump
(J) Flywheel puller
(K) Flywheel strap wrench
(L) Multimeter

General-purpose tools

(A) Socket set
(B) Standard screwdriver
(C) Phillips screwdriver
(D) Parts cleaning brush
(E) Power drill
(F) Putty knife
(G) Shot-filled mallet

(H) Center punch
(I) Flat file
(J) Needlenose pliers
(K) Adjustable pliers
(L) Star-shaped driver set
(M) Standard pliers
(N) Adjustable wrench

(O) Wire cutters
(P) Baster
(Q) Fuel line crimper
(R) Combination wrenches
(S) Hex wrench set
(T) Locking pliers

MORE SPECIALTY TOOLS

It's wise to have the correct tools for each job. And, this "right stuff" will save you time, too. Using other tools as substitutes can damage your engine, try your patience, and may create a safety hazard that'll have someone saying, "I told you so!" When buying tools, choose the highest quality you can afford. Shop around—from garage sales hosted by newly retired mechanically inclined folks who are downsizing, to big box home improvement stores offering decent discounts. And don't be afraid to ask an associate for details about the tool manufacturer. Well-made tools always pay off by helping you do a better job and they'll give you years of reliable service, too. The good ones are often worthy of being passed down to that next budding small-engine-buff in your family.

Tools you may want to purchase to simplify advanced projects include:

(A) Cylinder leakdown tester: for testing sealing capabilities of compression components

(B) Piston ring compressor: for compressing rings during assembly

(C) Starter clutch wrench: for removing and torquing rewind starter clutch

(D) Valve lapping tool: for resurfacing valve faces and seats

(E) Leakdown tester clamp

(F) Plug gauge: for checking valve guides for wear

(G) Telescoping gauge: for measuring inside diameters of cylinders

(H) Brake adjustment gauge: for setting band brake

(I) C-ring installation tool: for installing a starter motor c-ring

(J) Piston ring expander: for removing and installing piston rings

(K) Flywheel strap wrench: for removing and installing fly wheel

(L) Carburetor jet screwdrivers: for removing and installing carburetor jets

(M) C-ring removal tool: for removing c-ring on starter motor

(N) Torque wrench: for tightening bolts to specified "inch pounds" of torque

(O) DC shunt: for measuring current draw of DC motors and output of regulated alternators

(A) Moly graphite grease
(B) White lithium grease
(C) Carburetor/choke cleaner
(D) Fogging oil
(E) Heavy-duty silicone
(F) Battery cleaner
(G) Battery terminal protector
(H) 4-use lubricant
(I) Penetrating oil
(J) Heavy-duty degreaser
(K) Gasoline additive
(L) Valve guide lubricant
(M) Lawn mower oil
(N) Valve lapping compound
(O) Grease gun
(P) Tire sealant

SUPPLIES FOR SMALL-ENGINE WORK

To get the best results from your repairs and the highest performance from your small engine, use the lubricants and cleaners recommended by your authorized service dealer or outdoor power equipment retailer. Like tools, cleaners and lubricants are made for specific purposes, and each works best in the physical and chemical environment for which it was intended. Substituting one product for another could prove ineffective, damaging, or even dangerous. Your work will be easier and more reliable when you use the right tools, lubricants, and cleaners for the job. The lubricants and cleaners shown above represent the full range of products you need to keep your small engine in peak operating condition.

TIP

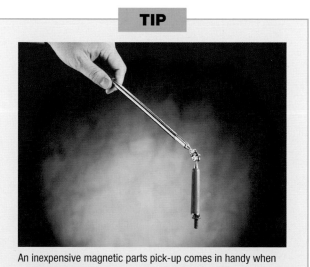

An inexpensive magnetic parts pick-up comes in handy when you drop small metal parts in hard-to-reach areas.

Safe Small-Engine Servicing

You can avoid a serious safety hazard—carbon monoxide (CO) accumulation—by working outdoors. CO is an odorless, tasteless, poisonous gas produced by burning gasoline and other fuels. Even an inexpensive carbon monoxide detector installed in your workshop/garage can alert you to the presence of CO indoors before it reaches lethal levels.

Small engines burn fuel and induce electricity. Each of these involves special safety considerations—so you need to observe the precautions for both. Keep the following rules in mind, and you will simplify the job of safely operating, maintaining and repairing your small engine.

OPERATING CONDITIONS

If you need to run an engine to test your maintenance or repair work:

- Never run an engine indoors.
- Turn off the engine before leaving the area—even for a few seconds.
- Do not operate the engine near combustible materials, gasoline, or other flammable liquids.
- Keep combustible materials away from the muffler and where you've got exposed spark (during an ignition test).
- Avoid running an engine at high speeds or in excess of the manufacturer's specifications.
- Make sure the muffler is in place before starting the engine.

- Pull the starter cord slowly until you feel resistance, then pull rapidly to start; this helps prevent injury to your hand and arm.
- Unless cautiously testing for spark, do not crank the engine with the spark plug removed; if the engine is flooded, place the throttle in the FAST position and crank until the engine starts.
- To stop the engine, gradually reduce engine speed. Then, turn the key to OFF or move the controls to the OFF or STOP position.
- When operating equipment on unimproved land covered by grass or brush, install a spark arrester—designed to trap sparks discharged from the engine.
- Keep equipment flat on the ground when it is operating. Never tilt it at a sharp angle.
- Keep hands and feet away from moving or rotating parts on the engine or equipment.

GASOLINE SAFETY

The only place where engine fuel and sparks should interact is in the combustion chamber. To reduce fire hazards:

- Never light a match or other flammable material near an engine.
- Avoid using power tools or other equipment that generates sparks where fuel vapors may be present.
- Allow the engine to cool before removing the fuel cap or filling the tank.
- Replace a fuel line or fitting if it is leaky or cracked.
- Keep gasoline, solvents, and other flammables out of reach of children. Store gasoline in UL®-approved non-spill containers. Label flammable materials containers clearly for quick identification. For 2-stroke engines, it's really important to label fuel containers to show the content's gas/oil ratio. Thinking you'll easily recall whether it's the ½ pint of oil-to-a-gallon container or one with the remnants of last season's ethanol-regular isn't always a winning guessing game.

SAFE MAINTENANCE

To make small-engine maintenance and repair tasks easier and safer:

- Make sure you have ample work space, with easy access to the tools you need.
- Use the correct tools for each job.
- Keep an approved fire extinguisher in a familiar location near your work area.
- Learn engine shutoff procedures so you can respond quickly in an emergency.
- Disengage the cutting blade, wheels, or other equipment, if possible, before starting the engine.
- Disconnect the spark plug wire to prevent accidental starting when you are servicing the engine.
- Always disconnect the wire from the negative terminal when servicing an electric starter motor.
- Check that a spark plug or spark plug tester is attached to the engine before cranking.
- Avoid contact with hot engine parts, such as the muffler, cylinder head, or cooling fins.
- Never strike the flywheel with a hammer or hard object; it may cause the flywheel to shatter during operation.
- Make sure the air cleaner assembly and blower housing are in place before starting the engine.
- Remove any fuel from the tank and close the fuel shutoff valve before transporting an engine.
- Use only the original manufacturer's replacement parts; any other parts may damage the engine and create safety hazards.

- Keep engine speed settings within manufacturer specifications. Higher speeds can ruin the engine. For example, don't run a lawn mower engine anywhere near full throttle minus its load (blade).

PROTECTING YOUR HEALTH

Fire, electric shock and asphyxiation are not the only dangers when working with small engines. Take care to avoid long-term or sudden injury to your eyes, ears, lungs, feet, and back:

- Keep your feet, hands, and clothing away from moving engine and equipment components.
- Use eye protection when you work with engines or power tools. This precaution is often ignored as particularly inconvenient—until some small part or tool breaks loose and flies right into one's irreplaceable peepers.
- Wear ear protection to reduce the risk of gradual hearing loss from exposure to engine noise.
- Wear a face mask, if required, when working with chemicals.
- Wear specially designed gloves to protect against heat, harmful chemicals, and sharp objects, such as mower blades or snowblower tines being held fast during removal.
- Wear safety shoes to protect against falling objects; safety shoes have soles that won't deteriorate when exposed to gasoline or oil.
- Use proper lifting techniques and seek help with heavy lifting.

FILLING THE FUEL TANK

Spilled or dripping fuel can cause harm to you, your equipment, and the environment. You can reduce these occurrences by using a non-spill can for transporting fuel and refueling. When filling, place your fuel can or power equipment on the ground—away from appliances, heaters, and other sources of flame or heat. Never fill your can or refuel your equipment while either one is inside a trunk or on a truck bed. During transport, fuel should be in a secure, upright position and tightly sealed. Provide ample ventilation to prevent fumes from building up in the passenger compartment or trunk, where static electricity could ignite gas fumes.

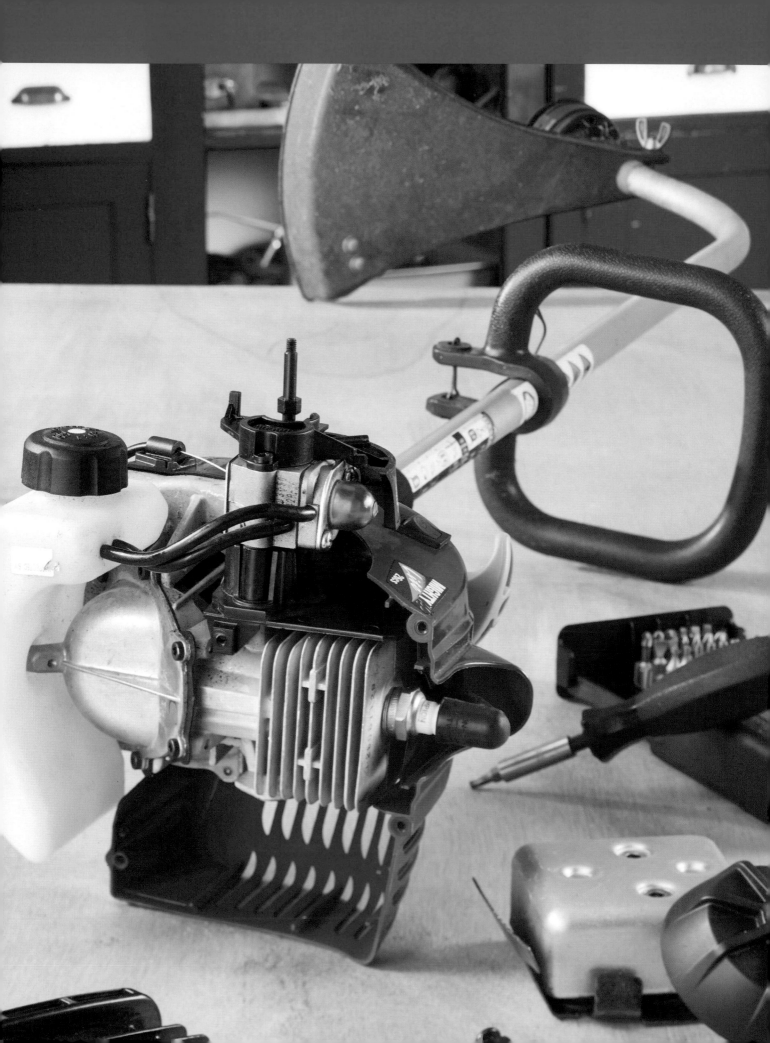

Motor Medical School 101

Sometimes the best way to explain information is with a story. This is as true of small-engine repair as it is true of anthropology or military history—well, maybe not as true, but my years of teaching have shown me that if you want to catch your students' attention, tell them a story. So here it goes:

Eric heard the motor on his recently purchased snowblower just kind of thin out, make some labored chugs, and then die. The back-saving but expensive piece of power equipment had only a few minutes on its clock when it went silent after clearing about eight feet of driveway. Frantically pushing on the fuel primer bulb and then yanking on the starter cord a dozen increasingly aggravating times did nothing but add to Eric's fast-growing fear that he'd been stuck with a lemon in the midst of a white-out blizzard. "What is the matter with this darn thing?" he yelled out into the cold air. There was plenty of gas in the tank, and the engine's little dipstick showed enough oil, too. A neighbor known for her mechanical acumen noticed Eric's dilemma. She walked toward the silent snowblower to take a look and try to help. Visually, she agreed, everything appeared to be in proper order. . . but nothing Eric consulted in the owner's manual and nothing the neighbor tinkered with produced a single pop. Even a test the neighbor conducted on the motor's ignition system ruled out problems with the sparkplug. Eric was ready to give up when his neighbor noticed something that caused her to shake her head and smile. Someone at the factory had misapplied the tiny fuel valve decal to the plastic shrouding in such a way that what looked to be ON was actually OFF. "Don't feel so bad," Eric's engine expert suggested told him when the quick fix had the engine running, "I almost missed that quick fix myself!"

—Peter Hunn

The details of this story and the specific problem they discovered are less important than the broader lesson learned: always check the simplest possibilities first.

Learning How to Diagnose a Motor's Symptoms

Small-engine repair projects should always start with this brand of troubleshooting—the search for the source of a problem—beginning with the most obvious or simple explanation and working toward the less obvious or more complex.

When you're troubleshooting a small-engine problem, you can gain confidence by being able to rule out, one by one through the process of elimination, various parts or systems as possible sources of the problem. For example, if your weed-whacker stops on its own after about twenty-seconds, even a novice mechanic could safely cross off the carrying handle as being the culprit; One down, several more to go. It's important to work systematically to isolate the cause rather than forever skipping parts or systems that have never given you any reason to believe are acting up. This forensic diagnosis process is a lot like looking for a lost set of keys: often, they're in an "obvious" place that didn't seem worth checking during much of the time spent in that frantic search. The solution is not to overlook things that may otherwise seem too easy a fix to be true. They just might hold the keys to enjoying a newly recuperated engine.

THE "DIAGNOSTIC TRIPLET" QUESTIONS:

Are you hearing strange sounds? Is anything leaking? Are there any smoky clues? The three queries above should be considered before you ever reach for a screwdriver or wrench.

Are you hearing strange sounds? If the engine can start, does it make any unusual noises? While there's no need trying to mimic a funny clang or knock, it is helpful to begin recognizing sounds that will help you pinpoint a mechanical problem. If the motor is sadly silent and absolutely nothing moves when you attempt to engage the starter, that sometimes indicates a serious technical difficulty, but not always. So, don't start worrying, yet.

Is anything leaking? Look for leaks on, under, and around the engine and its power equipment host. Fuel, oil, grease, and water are the usual suspects here. Note where they're coming from and whether or not they're still flowing or oozing. Discovering the "headwaters" source of a leak, and not just where the collective drips ended up, can be an invaluable troubleshooting breakthrough. Also inspect visually for cracks in the hoses and metal. Quite possibly, a loose lamp or fitting is allowing the leak.

Are there any smoky clues? If you are able to operate the engine, is it belching any unusual quantities or particular colors of smoke? Does the rate of speed increase or lessen this smoke? Is it worse or clearer *after* the engine warms up for a few minutes? To match your answers with possible causes, see page 40.

- Look for the overall cause, not just a temporary cure for the symptoms.
- Gather and jot down as much information as possible. Knowing whether an engine was at top speed when it stopped running or whether it simply failed to start may make a difference when you're trying to identify the problem. Remember the "diagnostic triplet questions" above. Also realize that a simple solution might not always bring the motor back to factory freshness or solve the engine's overall health issues, some of which will be clearer after arriving at one of the difficulties. For example, replacing a bad spark plug may get an engine running again, but the real culprit may be a dirty carburetor orifice for which the new plug was able to sufficiently compensate in order to get the motor firing for a while anyway.

Problem: Engine won't start

Fuel line problems—For more information, go to page 38

Carburetor problems—For more information, go to page 38

Ignition problems—For more information, go to page 39

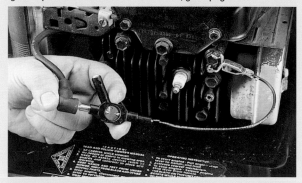

Compression problems—For more information, go to page 39

Problem: Engine runs poorly

Engine smokes—For more information, go to page 40

Engine overheats—For more information, go to page 40

Engine knocks—For more information, go to page 41

Spark plug misses under load—For more information, go to page 41

MOTOR MEDICAL
SCHOOL 101

TROUBLESHOOTING GUIDE

If this is the problem:	*Ask this question:*	*If the answer is yes:*
Engine/flywheel won't rotate 	*Is the starter jammed or rewind broken?*	Remove rewind/shroud and try rotating flywheel. If it does so freely, the rewind needs attention.
	Is the mower blade (or equivalent) hitting something?	Free the offending obstruction.
	Is the engine or blade brake activated?	Disengage it.
	Is an engine bearing seized?	Typically, the engine must be disassembled to determine this.
	Is the piston stuck in the cylinder?	A piston "stuck" from lengthy sitting can often be coaxed free by squirting the cylinder—through the spark plug hole—with penetrating oil or kerosene, letting it work for a day or so, and then moving the flywheel bit by bit, until the piston freely moves within the cylinder.
	Is the crankshaft bent?	This requires major disassembly and attention by a well-equipped shop.
Engine won't start (Fuel line) 	*Is the fuel tank empty?*	Fill the fuel tank; if the engine is still hot, wait until it has cooled before filling the tank (see "How to Remove and Clean the Fuel Tank," page 74).
	Is the shut-off valve closed?	Open the fuel shut-off valve (see "How to Remove and Clean the Fuel Tank," page 74).
	Is the fuel diluted with water?	Empty the tank, replace the fuel and check for leaks in the fuel tank cap (see "How to Remove and Clean the Fuel Tank," page 74).
	Is the fuel line or inlet screen blocked?	Disconnect the inlet screen from the engine and clean it, using compressed air. NOTE: Do not use compressed air near the engine (see "How to Remove and Clean the Fuel Tank," page 74).
	Is the fuel tank cap clogged or unvented?	Make sure the cap is vented and that air holes are not clogged (see "How to Remove and Clean the Fuel Tank," page 74).
(Carburetor) 	*Is the carburetor blocked?*	Remove the spark plug lead and spark plug; pour a teaspoon of fuel directly into the cylinder; reinsert the spark plug and lead; start the engine; if it runs for a moment before quitting, overhaul the carburetor (see Adjusting the Carburetor," pages 82 and 100).
	Is the engine flooded?	Adjust the float in the fuel bowl, if adjustable; make sure the choke is not set too high (see information on adjusting or overhauling the carburetor on pages 82 and 100).

If this is the problem:	Ask this question:	If the answer is yes:
Engine won't start (Ignition) 	*Is the spark plug fouled?*	Remove the spark plug; clean the contacts or replace the plug (see "Servicing Spark Plugs," pages 52 to 53).
	Is the spark plug gap set incorrectly?	Remove the spark plug; reset the gap (see "Servicing Spark Plugs," pages 52 to 53).
	Is the spark plug lead faulty?	Test the lead with a spark tester, then test the engine (see "Servicing Spark Plugs," pages 52 to 53).
	Is the kill switch shorted?	Repair or replace the kill switch (see "Servicing the Brake," pages 132 to 137).
	Is the flywheel key damaged?	Replace the flywheel key, re-torque the flywheel nut to proper specifications, then try to start the engine; if it still won't start, check the ignition armature, wire connections or, in some engines, the points (see "Replacing the Ignition," pages 110 to 113).
(Compression) 	*In a **4-stroke,** are the valves, piston, cylinder or connecting rod damaged?*	Perform a compression test (On any small engine, compression gauge should read 80 psi or more when the starter cord is pulled vigorously; if the test indicates poor compression, inspect the valves, piston and cylinder for damage and repair them as needed (see "Removing Carbon Deposits," pages 118 to 121, and "Servicing the Valves," pages 122 to 131).
	*If a **2-stroke,** are all of the crankcase seals intact?*	If no, replace seals.
	Is the reed valve working properly?	If no, replace reed valve.
	Any loose bolts on crankcase, carburetor mount?	Tighten bolts.
	Any bad gaskets allowing air in/out?	Replace gasket.

continued

MOTOR MEDICAL SCHOOL 101

Engine runs poorly
(Engine smokes)

	Is the smoke white?	Check for water (or coolant, if radiator-equipped) in fuel tank, line, carburetor, or cylinder.
	Is the smoke blue?	Indicates excessive oil is being burned. Check to see if piston rings are worn or stuck. Possible valve issue on 4-stroke. Possibly too much oil mixed in the gasoline, if 2-stroke engine.
	Is the smoke black?	A sign that engine is running rich (too much fuel in the fuel/air ratio). Carburetor needs adjusting.
	Is the fuel mixture too rich?	Adjust the carburetor (see "Adjusting the Carburetor," page 82).
	Is the air filter plugged?	Clean or replace the air cleaner (see "Servicing Air Cleaners," pages 54 to 57).

(Engine overheats)

	Is the engine dirty?	Clean the engine (see "Removing Debris," pages 69 to 71).
	Is the oil level low (in a 4-stroke engine crankcase)?	Add oil to the engine oil fill. NOTE: Never add oil to the gasoline for a 4-stroke engine (see "Checking & Changing Oil," pages 48 to 51).
	Is there too little, or excessive oil mixed with the gas in the 2-stroke engine?	Dump existing fuel and reintroduce a fresh gas/2-cycle oil mix, according to the manufacturer's specifications.
	Are any shrouds or cooling fins missing or broken?	Install new parts as needed (see additional tools, parts, and supplies, pages 28 to 31).
	Is the fuel mixture too lean?	Adjust the carburetor (see "Adjusting the Carburetor," page 82).
	Is there a leaky gasket?	Replace the gasket (see "Overhauling the Carburetor," page 100).
	On a 2-stroke, is there any crankcase leakage?	Check seals, gasket condition, tighten bolts.
	Is the fuel tank vent or fuel tank screen plugged?	Clean the fuel tank vent and fuel tank screen (see "Removing and Servicing the Fuel Tank," pages 73 to 74).

If this is the problem:	*Ask this question:*	*If the answer is yes:*

Engine runs poorly
(Engine knocks)

Does the noise get louder (as the motor warms up) and change pitch when the engine speed is increased?

This is probably a connecting rod knock requiring new connecting rod/rod end cap and/or bearings.

Is there a quick double knock especially noticeable at slow speeds?

Here, a loosely fitting piston pin (holding the piston onto the top of the connecting rod) is the likely suspect. It could also signal excessive wear in the piston pin hole or a loosely fitting connecting rod. Replacement parts are needed.

Is there a knock that is noisier than a rod knock, but seems to soften as the motor runs a while?

Perhaps "piston slap" is the issue. A worn piston ring, cylinder wall, and/or piston is allowing the piston to fit in the cylinder more loosely than designed for optimum clearance/compression. Replacement parts are needed.

Does the combustion chamber contain excess carbon?

Clean carbon from the piston and head (see "Removing Carbon Deposits," pages 118 to 121).

Is the flywheel loose?

Inspect the flywheel and key; replace as needed (see "Inspecting the Flywheel & Key," pages 108 to 109).

(Spark plug misses under load)

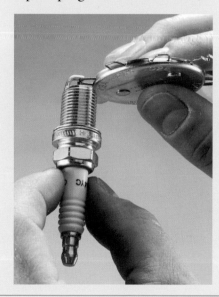

Is the spark plug fouled?

Clean the spark plug (see "Cleaning and Inspecting a Spark Plug," page 52–53). If plug looks/works OK, check condition of ignition wiring, especially plug lead wire and connection at end of lead wire where it attaches to spark plug.

Is the spark plug faulty or gap incorrect?

Replace the spark plug or adjust the spark plug gap (see "Cleaning and Inspecting a Spark Plug," page 52–53).

Are the breaker points faulty?

Install a solid-state ignition (see "Replacing the Ignition," pages 110 to 113).

Is the carburetor set incorrectly?

Adjust the carburetor (see "Adjusting the Carburetor," page 82).

*On a **4-stroke engine,** is the valve spring weak?*

Replace the valve spring (see "Servicing the Valves," pages 122 to 131).

*On a **4-stroke engine,** is the valve clearance set incorrectly?*

Adjust the valve clearance to recommended settings (see "Servicing the Valves," pages 122 to 131).

Easy But Important Maintenance

Our small-engine repair story continues (see page 35).

Eric always thought it rather curious that the guy across the street cut grass with an ancient 2-stroke lawn mower from the Kennedy era. After all, the fellow and his wife didn't appear to be fiscally downtrodden. When the couple walked past his driveway one early summer evening pushing a rusty, squeaky, rattly thing with just enough decal remaining to identify it as a LAWN-BOY mower, Eric couldn't resist engaging the duo in some investigative conversation. Before Eric had finished an introductory wave, the guy pointed to his smiling spouse and apologized, "Sorry for disturbing the neighborhood peace! My motorhead wife here saw this mower at somebody's curb, and decided to take pity on the poor old thing; Like other women might do for a stray puppy." She agreed and noted her latest acquisition was even more vintage than the one she'd rescued and restored and now used.

"Where'd you learn how to fix these things?" Eric asked.

"I guess you could say I was recruited into a training program by a small-engine shop," she smiled; "A very small small-engine shop. The owner taught me that anyone can repair small engines as long as she—or he—has a system and is disciplined enough to follow it. Anyway, most people only run their power equipment for an hour or two a week, and then for only part of the year. I guess you could say that the average small engine suffers much more from neglect than from being used."

—P. H.

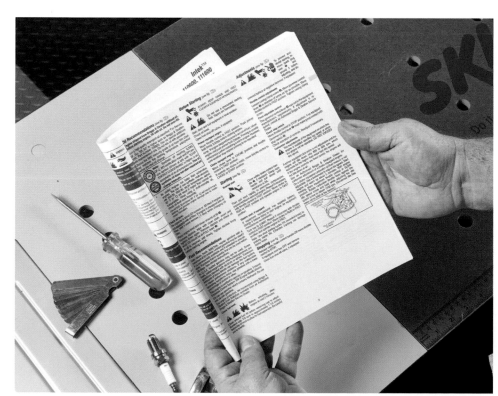

It is always a good idea to read through the owner's manual cover to cover when you acquire a new piece of equipment. But be sure to keep it handy in a convenient spot for times when you need to refer to it for specific information. If you lose your manual, or never had one in the first place, there are many websites where you can search for, locate, and download electronic files of the original owner's manuals—sometimes for no cost.

REGULAR MAINTENANCE SCHEDULE

You can avoid many small engine problems and save money on parts, repairs, and/or a pricey new piece of power equipment if you follow a regular maintenance schedule. Make routine maintenance a habit when your engine is new and always consult your owner's manual for special guidelines for your particular motor. It's smart to service your engine more frequently if you use it heavily or under dusty or dirty conditions.

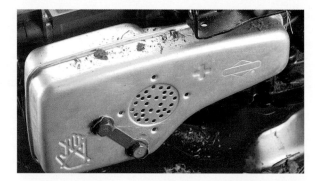

AFTER THE FIRST FIVE HOURS OF USE:

- Change the oil and filter (pages 48 to 51).

AFTER EACH USE:

- Check the oil (page 49).
- Remove debris from cylinder cooling fins, as well as around the fuel tank and muffler (pages 69, 70, 74, 94, and 95).

EVERY 25 HOURS OR EVERY SEASON:

- Change the oil if operating under heavy load or in hot weather (pages 48 to 51).
- Service the air cleaner assembly (pages 54 to 57).
- Clean the fuel tank and line (page 74).
- Clean the carburetor float bowl, if equipped (page 82).
- Inspect the rewind rope for wear (pages 90 to 91).
- Clean the cooling fins on the engine block (pages 69 to 71).
- Remove debris from the blower housing (pages 69 to 71).
- Check engine compression (page 58).
- Inspect governor springs and linkages (pages 86 to 89).
- Inspect ignition armature and wires (pages 110 to 113).
- Inspect the muffler (pages 92 to 95).
- Check the valve tappet clearances (pages 126 to 127).
- Replace the spark plug (pages 52 to 53).
- Adjust the carburetor (page 82).
- Check the engine mounting bolts/nuts (page 59).

EVERY 100 HOURS OR EVERY SEASON:

- Thoroughly clean the cooling system (pages 69 to 71).
- Change the oil filter, if equipped (page 50).
- Decarbonize the cylinder head (pages 118 to 121).

* Clean more often if the engine operates under dusty conditions or in tall, dry grass.

End-of-season maintenance

Engines are built to run. But many small engines are only used seasonally and sit idle for long periods. Long-term storage can aggravate overlooked problems, and other problems can develop. For example, unstabilized gas left in an engine can gum up a carburetor, unlubricated engine parts can corrode, and moisture can accumulate in the ignition system. With proper storage preparation, you can avoid most such problems.

If you plan to store your engine for more than thirty days:

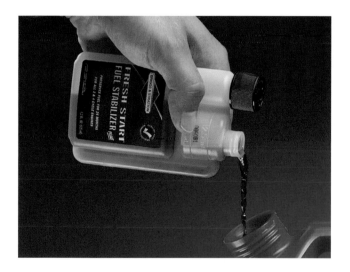

Drain the gasoline Gasoline that's allowed to stand for over a month may form a varnish on the inside of the fuel tank, carburetor and other fuel system components. Draining the gasoline reduces varnish problems. Another way to protect against the varnishing effects of old fuel is to add a gasoline stabilizer to your fuel, before storage (page 45). When you're done with your small engine's final chore for the season and are ready to turn it off, do so by closing the fuel valve and running it out of gas. Drain the carburetor float bowl (if equipped) as well. The Environmental Protection Agency recommends adding the drained gasoline to your car's gas tank, provided your car tank is fairly full. Once diluted, old gasoline will not harm your car engine.

Change the oil (4-Stroke only) Changing the oil will prevent particles of dirt in the oil from adhering to engine parts (pages 48 to 51).

Seal the fuel cap Your small engine emits small amounts of fuel vapor into the air—even when it's not running. To reduce emissions during storage, cover the vented fuel cap with aluminum foil and secure it with an elastic band.

End-of-season maintenance
(continued)

Lubricate internal parts Injecting oil through the spark plug hole is an easy way to lubricate the cylinder. Just squirt about ½ ounce of fogging oil (or some motor oil) into the spark plug hole. Then, spread it throughout the cylinder by reattaching the spark plug and very slowly pulling the rewind.

Inspect the spark plug Clean and regap the spark plug or replace it, as necessary (pages 52 to 53).

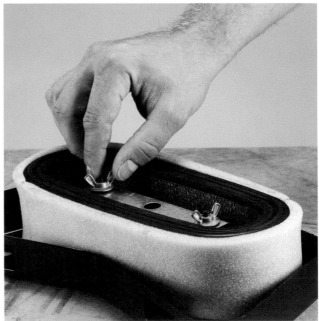

Service the air cleaner This step can extend the life of the air cleaner and improve engine performance next season (pages 54 to 57).

Seal the combustion chamber
You can prevent varnish formation in the combustion chamber during storage by placing the piston at top dead center (TDC), the point at which both valves (in a 4-stroke engine) are closed. This keeps out stale fuel and debris. Just pull the rewind rope slowly. When you feel increased tension on the rope (due to the compression of air in the chamber), the piston is at TDC (page 14). If your machine has a 2-stroke engine, slowly pull the starter cord until you feel the compression increasing, indicating the piston has covered the cylinder ports.

Remove dirt and debris then cover Debris tends to accumulate in the cylinder head fins, under the blower housing and around the muffler. This debris can fall into the engine. Remove it now to ensure good performance next season. Then, cover the engine with a sheet of plastic and store in a dry place. Be aware that critters, such as mice, get chilly and look for small openings (on mufflers, carburetor air intakes, and under shrouding) in which to build their nests. Their waste can really make a corrosive mess! Check for new "tenants" periodically.

Checking & Changing Oil on a 4-Stroke

Collect used engine oil in a plastic container that can be sealed securely and then dispose of it promptly, usually by bringing it to a service station or to a recycling facility that accepts oil.

Tools & Materials

- Socket wrench set, box wrench, or adjustable wrench

- Screwdriver or hex key

- Oil filter or pipe wrench (for models with filters)

- Oil drain pan

- Funnel

Time required: 30 minutes

When you pour fresh oil into the crankcase, it's a golden or amber color. Gradually, the heat, dirt particles, and agitated air in the crankcase cause the oil to darken. Dark oil is not only dirty; it has also lost much of its ability to coat and protect engine components. Manufacturers recommend changing the oil in your small engine after every 25 hours of operation. For a new engine, you'll also need to change the oil after the first five hours of operation. New engines require this extra step to flush out small particles that accumulate naturally during the break-in period.

Hours of use are just one factor in determining how often the oil should be changed; the amount of wear and tear is equally important. Just like the oil in a pickup truck operated in extremely dirty or dusty conditions or at high speeds, the oil in a lawn mower or other small engine breaks down faster under tough conditions, such as wet grass, heavy dust, high temperatures, and rough or hilly terrain.

Do not overfill your crankcase. Too much oil can cause the same type of engine damage as not having enough. Air bubbles form in the oil, reducing overall lubrication. The resulting friction and metal-to-metal contact can cause premature part failure. Excess oil can also burn in the cylinder, producing smoke and leaving carbon deposits.

This section covers procedures for checking and changing oil and oil filters, and offers tips on avoiding spills and choosing the right oil and other products for your engine.

Getting the old oil out of you outdoor power equipment can be tricky, especially with tools like lawn mowers where you have to tip the machine to get access to the drain plug. Try to do this when the gas tank is empty or nearly empty to minimize loss of fuel through the vent in the gas cap. And to maximize the amount of particulate matter you remove with the old oil, run the engine for a couple of minutes right before draining the oil so debris does not settle.

EASY BUT IMPORTANT MAINTENANCE

48

CHECKING THE OIL ON A 4-STROKE ENGINE

Make it a habit to check the oil level and appearance each time you're about to start a small engine. Checking the oil while the engine is cold and most of the oil is in the crankcase yields the most accurate reading. You won't need to change or add oil every time. But you'll ensure a better-running engine and avoid problems down the road if you keep the crankcase full and change the oil on schedule and any time the oil loses its amber hue. More than a few oil fill/drain caps have vibrated off of mowers and snowblowers without prompt owner detection.

How to check the oil on a 4-stroke

1. Start by locating the oil fill cap on the crankcase. Fill cap locations vary, depending on the make and model of your engine. On newer models, look for an oil-can symbol or the word "oil" or "fill" stamped on the plug. On small tractors, you may have to lift the hood to locate this cap. Some engines contain either an extended oil fill tube or a standard fill hole with a dipstick for inspection. Others require you to remove the fill cap to check that the oil is at the fill line or the top of the fill hole.

2. To prevent dirt and debris from falling into the crankcase, wipe the area around the cap with a clean cloth before removing the cap. If there is no dipstick, dab the oil with a clean tip of the cloth to inspect the oil.

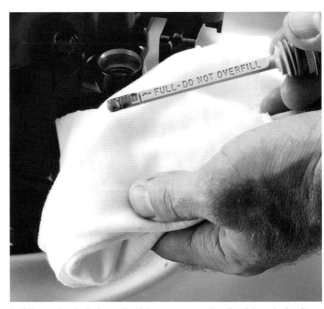

3. If the engine includes a dipstick cap, remove the dipstick and wipe it with a clean cloth. To ensure an accurate reading, reinsert the dipstick completely. Then, remove it again and check the oil level. If the dipstick cap is a screw-in type, ensure an accurate reading by screwing it in all the way before removing it a second time to check the level. The oil mark on the dipstick should be between the lines shown on the dipstick. It should never be above the FULL line or below the ADD line.

CHANGING THE OIL ON A 4-STROKE ENGINE

Once you decide the oil needs changing, check your owner's manual to determine the type of oil, and make sure you have enough on hand. Then, run the engine for several minutes. Draining the oil while it's warm will carry off many floating particles that would otherwise settle in the engine.

How to change the oil on a 4-stroke

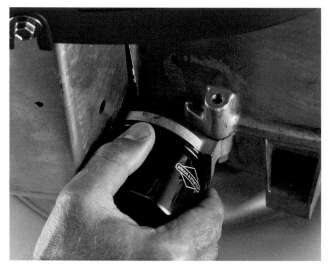

2. If your engine has a filter, replace the filter at least once a season, more often under heavy use (see "Regular Maintenance," page 44). Replace the filter by twisting counter clockwise on the body, using a filter wrench or pipe wrench. Lightly oil the filter gasket with clean engine oil. Install a new filter rated for your engine. Screw in the filter by hand until the gasket contacts the filter adapter. Tighten the filter an additional ½ to ¾ turn.

1. Stop the engine, disconnect the spark plug lead, and secure it away from the spark plug. Tilt the mower deck and position some newspaper and an oil pan or jug beneath the mower. Then, locate the oil drain plug (photo below). On mowers, the plug is typically below the deck and may be obscured by a layer of grass and debris. Wipe the area with a rag to prevent debris from falling into the crankcase when you open the drain plug. Use a socket wrench to turn the plug counterclockwise, allowing the old oil to drain. If the plug also serves as a fill cap, it may have two prongs so you can loosen it by hand or use a screwdriver or hex key for additional torque. Replace the drain plug by twisting clockwise and tightening with a box wrench or adjustable wrench.

3. Add the appropriate quantity of oil (see your owner's manual). Then, run the engine at idle and check for leaks. After an oil change, dispose of oil and soiled rags in accordance with local environmental statutes. In many areas, oil can be left at curbside with other recyclables, provided it is sealed in a recyclable container. Check the regulations in your area.

Drain plug

A few of the more popular grades of engine oil formulated especially for small engines. See descriptions below.

CHOOSE THE PROPER OIL

(A) 100 percent Synthetic Oil. Uniquely blended for use in all 4-stroke engines, all-season oil provides the extra protection needed in the most severe operating conditions. It exhibits excellent cold-weather flow characteristics that minimize engine wear and provide easier cold starting.

(B) 2-cycle engine oil. Product shown contains fuel stabilizer, as it is added directly to the fuel.

(C) Marine-grade 2-cycle engine oil is used with outboards, motorcycles, snowmobiles and other outdoor power equipment that has a 2-stroke engine.

(D) Small-engine oil is formulated to lubricate all 4-cycles engines, including those for lawn mowers, generators, pressure washers and other types of outdoor power equipment.

TIP

An *automotive funnel has an extra-long neck to make it easier to deliver fresh oil to your small engine without any spillage.*

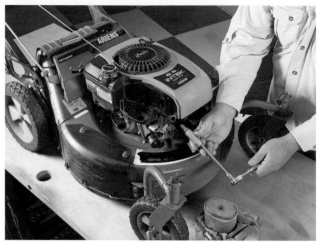

A spark plug socket is required for tightening and loosening spark plugs, but once they are spinning you can usually remove them by hand. Spark plug sizes and ratings vary quite a bit—be sure the plugs you buy are the correct ones for your engine—most parts dealers have charts and spec books with this information.

 Tools & Materials

- Spark tester
- Spark plug socket (sizes vary)
- Socket wrench
- Wire brush
- Plug/point cleaner
- Spark plug gauge

Time required: 15 minutes

Servicing Spark Plugs

The electrodes on a spark plug must be clean and sharp to produce the powerful spark required for ignition. The more worn or dirty a spark plug, the more voltage—and the greater the tug on the rewind—required to produce an adequate spark.

If you haven't tuned your engine recently and have to yank repeatedly on the rewind to start the engine, a damaged spark plug may be the culprit. Inconsistent firing, known as spark "miss," can result in sluggish engine operation and poor acceleration. A damaged spark plug may also cause excessive fuel consumption, deposits on the cylinder head, and oil dilution.

Luckily, a spark plug is one of the easiest engine components to repair and an inexpensive one to replace. And your standard socket set may already include the most important tool—a spark plug wrench.

This section covers the essentials of spark plug inspection and replacement. It shows you how to use a spark tester and how to adjust and clean a spark plug that is worn but still serviceable. Just remember, you can't go wrong by replacing it.

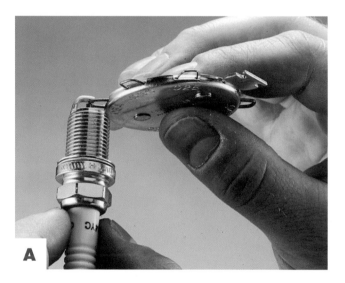

A

CLEANING AND INSPECTING A SPARK PLUG

1. Disconnect the spark plug lead. Then, clean the area around the spark plug to avoid getting debris in the combustion chamber when you remove the plug.
2. Remove the spark plug using a spark plug socket.
3. Clean light deposits from the plug with a wire brush and spray-on plug cleaner. Then, use a sturdy knife if necessary to scrape off tough deposits. NOTE: Never clean a spark plug with a shot blaster or abrasives.

4. Inspect the spark plug for very stubborn deposits, or for cracked porcelain or electrodes that have been burned away. If any of these conditions exists, replace the spark plug.

5. Use a spark plug gauge to measure the gap between the two electrodes (one straight, one curved) at the tip of your spark plug (photo A). Many small engines require a .030 inch gap.

Check the specifications for your model with your power equipment dealer. If necessary, use a spark plug gauge to adjust the gap by gently bending the curved electrode. When the gap is correct, the gauge will drag slightly as you pull it through the gap.

6. Reinstall the plug, taking care not to overtighten. Then, attach the spark plug lead.

CHECKING IGNITION WITH A SPARK TESTER

A spark tester offers an inexpensive, easy way to diagnose ignition problems (see "Checking for Spark Miss," below).

If you find a problem, remove and inspect the spark plug. Replace the spark plug if you find evidence of wear or burning at the spark plug tip. Spark plugs are inexpensive and a new one may solve the problem.

1. Connect the spark plug lead to the long terminal of your tester and ground the tester to the engine with the tester's alligator clip.

2. Use the rewind or electric starter to crank the engine, and look for a spark in the tester's window.

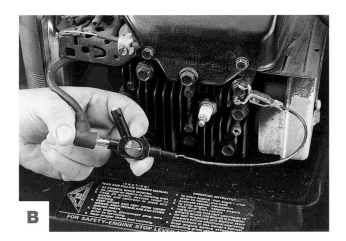

3. If you see a spark jump the gap in the tester, the ignition is functioning. The absence of a visible spark indicates a problem in the ignition system.

CHECKING FOR SPARK MISS

A spark plug that is fouled or improperly gapped may not allow sparks to jump the gap between electrodes consistently. The spark plug will fire erratically or may occasionally fail to spark. Test for this problem—known as spark "miss"—if your engine stumbles, with a noticeable decrease in engine sound. Spark miss can also cause the engine to emit black smoke or a popping sound, as unburned fuel exits with the exhaust and ignites inside the muffler.

1. With the spark pug screwed into the cylinder head, attach the spark plug lead to the long terminal of the spark tester. Attach the tster's

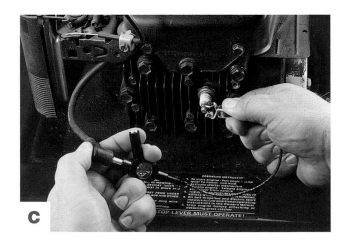

alligator clip to the spark plug (photo C).

2. Start the engine and watch the tester's spark gap. You'll recognize spark miss by the uneven timing of the sparks in the tester.

Tools & Materials

- Engine oil
- Screwdrivers
- Liquid detergent

Time required: 30 minutes

The air cleaner filters intake air on its way to the combustion chamber and must be cleaned or replaced when it becomes dirty or clogged.

Servicing Air Cleaners

A properly maintained air cleaner is your engine's first line of defense against the destructive effects of dirt. When the air cleaner is in good condition, it keeps airborne dirt particles from entering through the carburetor. If the air cleaner is not maintained, dirt and dust will gradually make their way into the engine. And don't underestimate dirt's potential to cause damage. It can lead to a sharp drop in engine power, or, worse, cause premature wear of critical engine components.

Many types of air cleaners are used in small engines. Most contain a foam or pleated-paper element.

Dual-element air cleaners contain a pleated-paper element with a foam precleaner, offering two layers of protection. Discard the paper element when you can no longer remove dirt from the pleats by tapping the element on a hard, dry surface. You may be able to wash and reuse the foam precleaner. Foam elements can be cleaned with hot water and liquid dish detergent that contains a grease-cutting agent.

Single-element air cleaners should be serviced every 25 hours (or once a season). In a dual-element system, the precleaner should be cleaned every 25 hours. The cartridge should be cleaned every 100 hours. Refer to the photo (opposite) to identify the air cleaner on your engine.

COMMON AIR CLEANERS

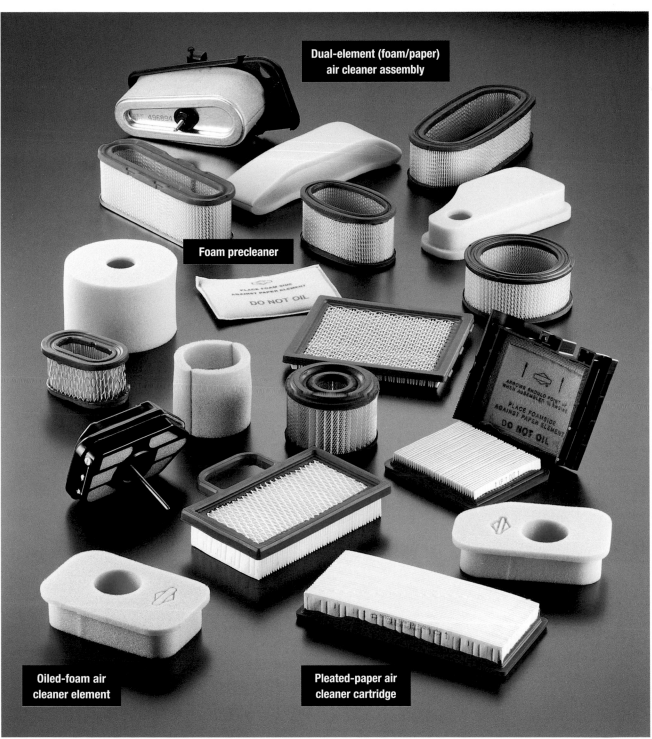

Dual-element (foam/paper) air cleaner assembly

Foam precleaner

DO NOT OIL

ARROWS SHOULD POINT UP WHEN ASSEMBLED IN ENGINE
PLACE FOAMSIDE AGAINST PAPER ELEMENT
DO NOT OIL

Oiled-foam air cleaner element

Pleated-paper air cleaner cartridge

Some small-engine air cleaners contain an oiled-foam element, others the pleated-paper cartridge. Newer engines often contain a combination of the two. Dual-element designs consist of a pleated-paper cartridge and a foam precleaner. Some precleaners are designed to be oiled.

A mesh backing separates the oil on the precleaner from the surface of the paper element. Others read "Do not oil." Identify the type of air cleaner on your engine before cleaning.

Don't wait until your air cleaner element looks like this to replace it. If your element has become permanently discolored or has begun to break down or tear, extend the life of your small engine by installing a new one.

How to Service a Foam Air Cleaner

1. Loosen the screws or wing nuts that hold the air cleaner assembly in place. Disassemble. Inspect the foam element. Replace it if it is torn or shows signs of considerable wear.

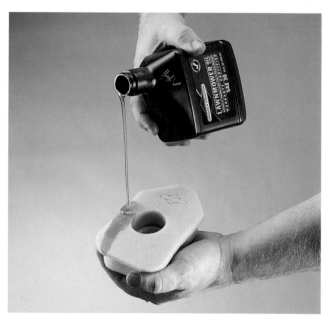

2. Saturate the new element with engine oil. Then, squeeze it to spread the oil throughout. Wrap foam in clean cloth and squeeze to remove excess oil. Inspect the rubbery sealing gasket between the air cleaner and carburetor. Replace it if it is worn. Reassemble and reinstall the air cleaner.

PAPER FILTERS AND TIPPING THE ENGINE

If your 4-stroke engine has a paper filter cartridge, remove it temporarily any time you are preparing to tip the engine on its side. You'll eliminate any chance that oil from the precleaner will spill onto the paper and ruin it. To prevent debris from entering the carburetor, temporarily cover the carburetor opening with plastic.

Pleats in a paper element that are discolored, bent or water-damaged can no longer provide adequate air to the carburetor. Replace the element when it approaches this condition.

How to Service a Pleated Paper or Dual-Element Air Cleaner

Dual-element air cleaners come in a variety of designs. Two of the most common are shown here.

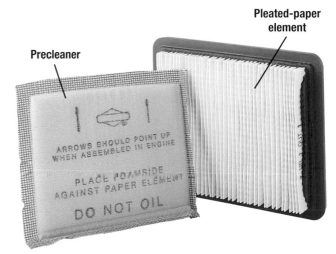

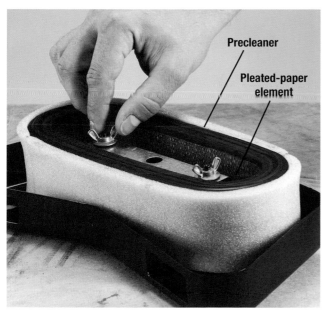

1. With the cover removed, separate the precleaner (if equipped) from the cartridge. Tap the cartridge gently on a flat surface to remove any loose dirt. Inspect the element and replace it if it is heavily soiled, wet, or crushed.

2. Inspect the precleaner, if equipped. Note the mesh backing, designed to act as a barrier between the oily precleaner and the pleated-paper element. If the mesh backing is plastic, you can wash the precleaner, wring it out, and let it dry. Don't wash a pre-cleaner containing a metal backing; replace it when soiled or worn. Look for oiling instructions on the precleaner. If directed, lubricate the precleaner with oil. NOTE: Not all foam precleaners should be oiled.

3. Clean the cartridge housing with a dry cloth. Do not clean with solvents or compressed air. Reassemble the air cleaner. If the pre-cleaner is the oiled type, take care to insert the mesh toward the paper element so that the paper is never exposed to the oil. Reinstall, making sure that any tabs on the cartridge are in their slots on the engine housing. Gaps around the cartridge permit unfiltered air and damaging dirt particles to enter the engine.

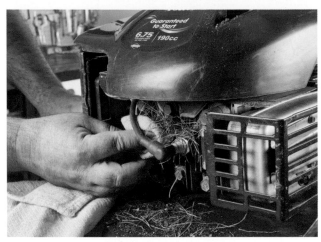

Cleaning off debris and gunk is a simple but critical part of any small-engine maintenance program.

More Smart Maintenance

This section covers four additional great ways to help ensure a nice-running engine. Testing engine compression, inspecting a crankcase breather, lubricating cables and linkages and tightening bolts. Take these steps when you're performing your preseason maintenance, or any time your engine and equipment have been operating under a heavy load or dirty or dusty conditions for a while.

Some more specialty tools that can come in handy when doing small-engine maintenance include: Extension wrench head (A); six-sided box wrench (B); large offset box wrench (C); ratchet wrench (D): adjustable box wrench (E).

TESTING ENGINE COMPRESSION

If your engine has leaks around the valves (4-stroke only) or rings, compression of the air-fuel mixture suffers. When this happens, performance and efficiency can drop dramatically.

A spin of the flywheel will give you a good idea of whether or not the compression in your engine is OK.

How to Test Engine Compression

Disconnect the spark plug lead and secure it away from the spark plug. Remove the blower housing. Then, disconnect the brake pad or band, if equipped (see "Servicing the Brake," pages 132 to 137). Spin the flywheel counterclockwise by hand. If compression is adequate, the flywheel should rebound sharply. A weak or nonexistent "bounce," indicates poor compression (see "Troubleshooting," page 39, for a list of possible causes and remedies).

OPTION: For a numeric compression reading, remove your small engine's spark plug, ground the plug lead to the engine block, and screw the hose fitting of a simple compression tester (available at most auto parts stores) into the spark plug hole. Pull vigorously on the starter cord (or engage the electric starter). Many experts say a reading of 100-psi represents the bottom of the low range for good starting/running, though some engines defy this decree by working OK with 80-psi. Your engine's service manual or local shop proprietor should help clarify a "healthy" number. In any event, knowing an engine's cylinder compression is as valuable in small engine diagnostic scenarios as is a doctor knowing his/her patient's blood pressure.

Tools & Materials

- Socket wrench set
- Needlenose pliers
- Feeler gauge
- Spray solvent/lubricant

Time required: 15–45 minutes

INSPECTING A CRANKCASE BREATHER

Many engines contain a crankcase breather to vent gases that accumulate in the crankcase. The breather (if equipped) is usually located over the valve chamber.

How to Inspect a Crankcase Breather

1. Remove the muffler or other parts to reach the breather. Then, loosen the breather retaining screws and remove the breather. Make sure the tiny holes in the body are open. Use a feeler gauge to check the gap between the fiber valve and the breather body. If a 0.045 in. feeler gauge can be inserted, replace the breather. Avoid using force or pressing on the fiber disc, and never disassemble the breather. If the breather is damaged, replace it. Replace the old gasket that fits between the breather and the engine body whenever you remove the breather.

LUBRICATING CABLES AND LINKAGES

Control cables and linkages on the governor, flywheel brake and throttle can seize and may even throw off engine performance if they can't be moved freely. You can reduce binding to a minimum and keep cables and linkages free of dirt and debris by spraying them occasionally with a solvent/lubricant, (photo, right).

TIGHTENING BOLTS

Bolts on your engine must be tight at all times. If the bolts remain loose, parts can easily be damaged during engine operation. Because mechanical designers understand an engine's stress and vibration can cause even "factory-tight" bolts to wiggle loose, they sometimes include "keeper" washers in their creations. Typically on the bolts securing components like a connecting rod to its rod cap, these washers are bent in one spot against its bolt (or nut) flat, as a way to prevent movement. Sometimes, the factory simply knocks a tiny "keeper" dimple into the aluminum where the bolt head (if in the form of a screw/hex head) is recessed. Trying to otherwise tighten or remove such "kept" bolts is later understandably maddening.

Mounting bolts that attach the engine to the equipment can also loosen, leading to damage, such as a cracked engine block. Check these and all other accessible nuts and bolts during regular maintenance, and any time you sense excess vibration.

Some mounting bolts must be grasped with a wrench from above and their associated nuts tightened with a second wrench from beneath the equipment to keep both ends from spinning. Others are self-tapping, and overtightening can damage the threads. Consult your service manual or local service shop for the proper torque for each bolt on your engine, and use a torque wrench for final tightening.

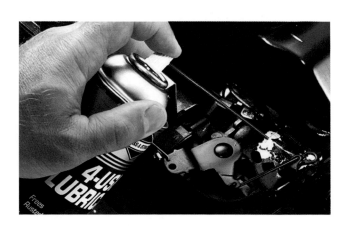

Use a rotary tool with a grinding wheel or a metal file to sharpen blades. The blade should be removed from the machine and secured in a shop vise.

Check the balance in your blade by hanging it at the center hole from a horizontal nail and observing if the blade top is level. Grind off metal along the blade edge to lighten the heavy side, if any.

- **Mowing Deck:** There's not a mower in existence that wouldn't appreciate the underside of its deck being cleaned of grass clippings and whatnot after every use. This is best accomplished after removing the spark plug lead (preventing inadvertent starting) and tipping the mower enough to gain access to the blade area without spilling fuel and (on 4-stroke-equipped units) causing oil to run out of the crankcase. With hands in sturdy garden gloves, and a trusty paint scraper in one hand, simply scrape away as much debris as possible from the deck and blade. A stiff wire brush can help, too. Some folks swear by the practice of using an old paintbrush to swab a coating of used motor oil onto the mower deck. All of this should result in a machine that lasts many seasons longer than it otherwise would.

- **Blades:** Remove the blade from the crankshaft by blocking the blade from turning via piece of wood and taking off the bolt(s) holding the blade in place. Put the blade in a vise, and while wearing eye protection, sharpen the blade by using either a flat file for metal use or a rotary tool outfitted with an appropriate sharpening wheel (which will work much faster). If you use a file, push it across the cutting edge of the blade using the original angle cut by the manufacturer. Filing works on the forward stroke while pushing it down with requisite force. Don't push back with the file over the blade. Aim for a nice shiny sharp surface with the prescribed angle. Those with the rotary tool or a bench grinder can get the same job done faster.

When satisfied with the blade sharpness, insert an ordinary nail sideways in between the slightly opened jaws of the vise and then close those jaws so that the nail is sticking out sideways, pointing parallel to the floor. Balance the blade (through the center hole) on the nail. If there's a heavy end, file that side's blade surface a bit more and aim for good balance. Such equilibrium is needed for keeping the mower from vibrating like crazy. Reattach the sharpened blade, set the mower upright, attach the spark plug lead, and go get that lawn!

- **Drive Units:** If you find that a so-called "self-propelled" mower is increasingly depending on you to push it, the drive system probably needs work. While gears may be employed, most of these machines utilize a simple drive belt transmission to shift from neutral to forward. Weak motion or lagging is typically the result of loose or worn belts. Grass clippings and moisture are usual suspects, too. Cleaning/tightening is the first order of business here. Admittedly, some gear-driven models suffer from problematic plastic gears. No matter the design, though, cables and linkages serving to engage the drives can be cleaned and better secured, thus facilitating improved forward mower motion.

TILLER MAINTENANCE TIPS

To most observers, all tillers look pretty much the same. It's probably the gangly, nonmobile appearance of garden machines or their reputation for being difficult to wheel that frequently causes them to end up in long-term storage with a bucket over their engine and in back of some weedy shed. Unless they are equipped with a transmission drive system that activates the wheels, they depend upon the forward motion of their churning or cutting tines for propulsion. That said, it is a wonder that old power tillers continue doing so much dirty work for their owners. If the engine is not an issue, tiller troubles are almost always belt and pulley related. In these models, cleaning, tightening, and possible belt replacement will improve performance. More deluxe examples have worm gear drives that require lubrication or an overhaul if the gears become stripped.

POWER AUGER MAINTENANCE TIPS

Essentially a small engine on a screw-type post-hole digger, ice augers are incredibly simple when compared to, say, a snowblower or garden tractor. Still, other than engine maladies, they are not immune to ailments, especially in the centrifugal clutch or other transmission mechanism that connects the motor to the auger blade. A high engine idle can start the auger turning immediately upon start-up. Adjusting the idle to a slower purr should solve this. If not, it's possible that the clutch spring or related parts might have come loose or broken, resulting in a jammed drive system. Here, the only way to determine this is to take it apart for inspection.

LAWN TRACTOR MAINTENANCE TIPS

The terms garden tractor and lawn tractor are often used interchangeably by the average individual, but the two machines—though appearing pretty much identical to that basic observer—are not the same thing. Typically, lawn tractors (sometimes dubbed "riding mowers") are primarily designed for cutting grass around homes with yards up to about an acre. There, maneuverability is key, economy is important, and parking in a modest space is appreciated.

So, a "garden variety" (so to speak) lawn tractor will probably have an engine rated at less than 15-horsepower, be gifted with sharp turning capabilities, wear smaller diameter but wider tires than a garden tractor, and sport a mowing deck cut width (with two or three blades) up to some 42 inches.

Garden tractors may look leaner and more farm tractor-esque than lawn tractors while getting powered by motors bumping into the upper range of the 25-horsepower maximum unofficial definition of a small engine. With this muscle and taller tires, a garden tractor can be equipped with a bigger mower deck or pull-behind gang mowers, and nicely handle attachments such as a trailer, grading blade, plow, or snowblower. Proper tire pressure is especially vital when adding these implements to a garden tractor or for any outdoor power equipment with pneumatic tires.

Some modern lawn tractors and garden tractors transfer power from engine to wheels via an automotive-type "stick-shift" manual gearbox or perhaps a hydrostatic transmission. The latter employs a variable pressure pump (typically enlivened by a belt drive off of the tractor's engine), pressure relief valve and pistons that send power to a hydraulic motor.

A control lever works in concert with the pistons and pressure valves to either increase or decrease the hydraulic fluid flow to the hydraulic motor, causes this motor to speed up or slow down. That means, as long as the engine is running at a sufficient rpm to easily activate the pump, the tractor's speed is essentially a province of the hydrostatic transmission.

While a manual transmission with ranges of meshing gears requires a reservoir of conventional oil in the gearbox and drive differential, one equipped with the hydrostatic transmission must have airtight lines and components able to withstand up to 375 psi. Such trannies get a lot hotter than a regular gearbox, so are fitted with a cooling fan.

The two big maintenance requirements in hydrostatic units are sufficient fluid (changed at least annually) and cleanliness of the transmission cooling system. Pushing or towing a hydrostatic transmission-equipped machine is not advised —due to possible fluid back pressure issues—unless otherwise allowed by the manufacturer.

There are still lots of lawn tractors in service that have belt, pulley, and friction wheel transmissions void of gears. Their primary maintenance includes cleanliness, and making sure that the belts, pulleys, and related control cables/linkages are properly engaging. Tracing the source of slippage or a funny noise, and then tightening the occasional loose pulley and/or replacing a frayed or broken belt keep this old technology attractively simple.

Lawn/garden tractor mower decks like the same maintenance of their smaller lawn mower sisters. The rigamarole of detaching these decks for cleaning access, however, can be frustrating when trying to negotiate a heavy cutting unit and its related belts and pins. It's best to make it a two-person project.

As in the lawn mower version, be sure to wear sturdy garden gloves when working around blades, and remember to first, either remove the ignition key or pull the spark plug wire off of its plug.

EASY BUT IMPORTANT MAINTENANCE

Routinely check all of the belts, including the propulsion and auger drive belts, for proper tension. The belts have a tendency to loosen.

Check the auger gear-case oil to make sure the levels in the reservoir are adequate.

Even with a beautifully performing engine, snowblowers need care to keep them happy in winter.

• Don't get snow in the oil fill spout, but make a habit of checking the lube level before each time the machine is started.

• Check the auger gear-case oil after every 25 hours of use, greasing the gears that help turn the snow-thrower chute, and checking drive belt tension and linkages after each five hours of operation.

• Keep the propulsion and auger drive belts properly tensioned, as wet loose belts slip easily, thus reducing the snowblower's effectiveness.

• On metal-to-metal surfaces, though, lubrication is vital to keep parts moving with agility. Your snowblower's manual should describe these needy points. It may give special attention to the grease fittings where the auger blade rides on its rotating shaft.

• A shear pin typically keeps the auger rotating with this drive shaft, but rust (from neglecting greasing) can cause the auger to bond to the shaft, preventing the auger from rotating freely if it hits a serious obstruction. And, that could result in all kinds of safety and mechanical problems.

• Veteran snowblower owners agree that the most valuable tip regarding operating such a machine safely is a personal pledge to never clear a snow clog from the auger or chute area by hand. Instead, use something like a long piece of 1-inch-square scrap lumber to coax away a clog. Be careful not to stand in the way of the chute while clearing the obstruction.

• All but the smallest snowblowers are of the two-stage variety, with the main auger as stage one and blower blade near the chute's base whirling rapidly to draw up the snow eaten by the auger and then blow it out of the chute. If either one of those components isn't turning properly, the other can't do its job.

FINALLY, A COUPLE OF GREAT WAYS TO CLEAN UP

Those who come to appreciate outdoor power equipment, or "OPE," as motor enthusiasts and small-engine professionals say, find themselves going in one of two OPE ownership directions; finding and enthusiastically revitalizing motorized equipment that has been given up by "regular folk" as hopeless; or, developing an eye for high-end, preowned equipment and finding great deals from well-heeled private sellers simply wanting the latest model. One such buff has filled his storage shed with these values, then mused on a small-engine blog that he loves top-of-the-line OPE and "appreciates the quality and performance that typically goes with it." But, he also admits being unwilling to fork over big bucks for premium products, which is why the guy buys almost all of his OPE as gently used equipment, thus saving significant money. To that end, he observes that most people who originally purchased high-end, high-priced OPE off a shiny showroom floor are usually dedicated to taking very good care of their equipment or routinely pay to have it serviced professionally. Because finances are not a particular issue with wealthier homeowners who decide to upgrade to something even neater and newer, they often quickly sell—to someone who'd truly appreciate it—their late-model mower, snowblower, or other OPE. Craigslist and fancy front yards are great places to spot those toney opportunities. On the opposite end of the spectrum, small-engine buffs wanting to practice the kind of repairs outlined in this book, will search curbs on trash day or look anyplace else harboring rusty stuff. There, one may realize a treasure trove of discarded OPE hoping to become part of somebody's enjoyably useful pastime!

Basic Engine Repairs

Our small-engine repair story continues (see pages 35 and 43).

There was an ancient little gas station less than a mile from Melinda's home. The old woman whose husband repaired lawn mowers, garden tractors, and the occasional outboard motor there, noticed Melinda struggling to keep her bicycle going. "Got a flat, Missy?" she called out. Melinda wobbled her bike to a stop. "Herb!" she yelled to her spouse, "This nice girl needs her tire fixed."

"Kiddo," he began with Melinda, "get me that bristle brush and rag over there. We gotta get things clean first. And we'll need the ⁹⁄₁₆-inch combination wrench and that Phillips screwdriver with the yellow handle on the workbench." Though at first she had almost no idea what he was talking about, his method of showing her his "six point plan," started making sense. "Number one—clean anything that prevents a good look," the old man started counting. "Two, inspect the situation. Diagnosing the trouble is number three. Then, four, prepare your tools. Step five is when you do the repair work. And finally, step six is testing your work to see if it was a success." Putting this six-step plan into action, the flat was repaired very quickly.

A few days later, when she returned to the gas station with a plate of thank-you cookies, Melinda got offered a Saturday job helping Herb and his wife around the shop. Wherever her assignment, she was schooled in the value of following some form of the old fellow's plan—from clean-up, triage-related tool selection, to working systematically on a repair project and then carefully testing the results. She also found it to be a methodical format translatable to schoolwork and useful in the engineering career she eventually pursued.

"If you can follow a simple system, you can fix just about any problem."

—P. H.

A workshop doesn't need to be a fancy place out of a cable TV gearhead show. Just a clean area in your garage with a mechanism for hanging tools and parts, some good light and ventilation, and space to work will do.

Basic Repairs

Making effective small-engine repairs starts with your workspace. It should be neat, well ventilated, well lit and well equipped. A sturdy workbench, vise, magnetic pick-up tool, air-compressor or bicycle pump, and broom can come in mighty handy during your repair projects. You can make a simple small-engine stand out of scrap lumber, allowing the power plant to be removed from its equipment mount, but a lawn mower or snowblower engine that is still mounted to its host is usually accessible and access is improved by positioning the power equipment on a large piece of cardboard or a tarp. That way, any dropped parts will have a better chance of being quickly found, and this work area can be more easily kept clean than might a bare garage floor or driveway. Taking digital pictures before each disassembly step will provide you with a vivid record of what went where. You might not have to refer back to the images, but if you ever do, they'll prove invaluable. Zip-top sandwich bags make great parts "keepers," especially when they're labeled with the removed part/system names. Also useful is an old board with some nails hammered in just enough to provide convenient hooks for washers, nuts, seals, gaskets, etc. If they're placed on the "keeper" nails in order of removal from the engine, reversing the process will be obvious. Finally, never underestimate the value of keeping your small-engine repair area clean. Being careful not to inadvertently discard parts (or clues in the form of broken pieces), periodic blowing/sweeping away of dirt, getting rid of greasy globs, and clearing formerly caked-on grassy grime lends itself to organization best practiced for successful work. That's why our first "warm-up" repair project involves the wise habit of keeping one's small engine clean and cool.

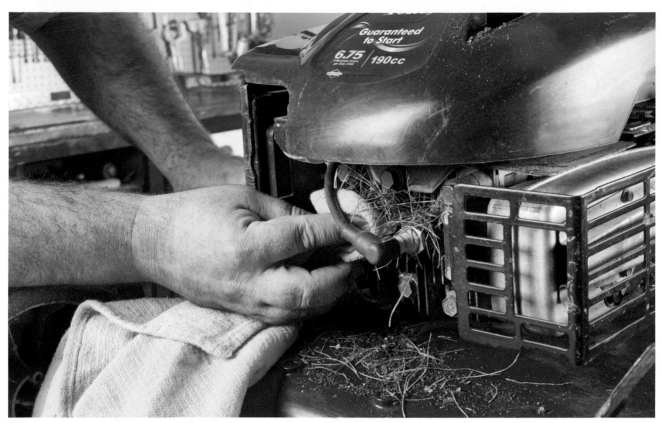

Completely removing debris from your small-engine systems is an ongoing battle that requires several weapons, from wiping cloths to compressed air.

Removing Debris

Grass and other debris may hardly seem like a critical repair issue for your small engine. But once it accumulates in between engine parts, it can cause a temporary loss of power or even permanent engine damage. Debris under the blower housing or in the cooling fins on the cylinder head can make an engine run too hot. Prolonged overheating may cause a piston to seize in its host cylinder. Debris can also cause governor linkages to bind or prevent air from reaching the governor blade on a pneumatic governor, resulting in difficulty controlling engine speed. Inspect the blower housing and muffler area for debris each time you use your engine. If the screen over the blower housing is clogged, it's a good indication that debris has accumulated underneath as well. Remove the blower housing for a more thorough inspection and cleaning at the end of each season of use and more often if you operate your equipment in tall or wet grass.

To fully remove all the debris buildup (the photo above shows a fairly advanced case) often means doing some light disassembly to gain full access to the areas where debris collects.

How to Inspect for Debris

1. Start by disconnecting the spark plug lead and securing it away from the spark plug. Snap off the plastic blower housing. If the housing is metal, you will need to remove a set of screws or bolts. On some models, removing the screws requires a star-shaped screwdriver or socket. A complete set of common sizes is available at most hardware stores. Clean the cooling fins, the inside of the blower housing, and the flywheel fins, using a small bristle brush.

2. Scrape dirt away gently, using a stiff bristle brush or a putty knife. Take care not to damage the housing or flywheel.

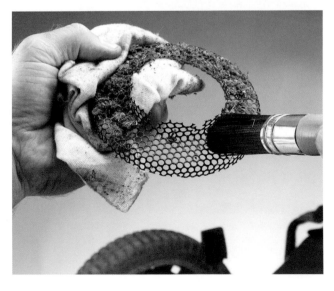

3. To loosen stubborn grit, apply a light solvent, such as kerosene, to the brush. Beware of spraying the likes of starting fluid or carburetor cleaner onto any parts where paint/decal removal is unwanted. Dirt and debris on the flywheel cutting screen can lessen the engine's ability to cool itself. Clean the screen thoroughly with a brush.

4. Remove all debris by hand or with the knife, screwdriver tip, and brush. Avoid using compressed air here, as to not force debris into less accessible engine parts. Remove any debris from governor linkages, including the pneumatic governor vane, if so equipped. Then, make certain linkages are moving freely, using a light solvent to loosen remaining dirt and debris.

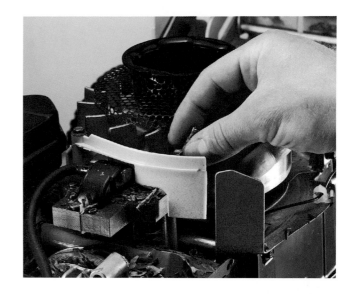

5. Check for debris around the brake assembly. Make sure the brake cable and linkage move freely.

6. Reattach the blower housing and reconnect the spark plug.

CHECKING THE STOP SWITCH

If your lawn mower stops unexpectedly while you're mowing around trees or bushes, you may have accidentally disconnected the stop switch wire, a short wire extending from the brake assembly to the ignition armature. A disconnected stop switch wire may ground the ignition, preventing the spark plug from firing. Under ordinary conditions, the stop switch is designed to stop the engine any time you release the brake bail on the equipment handle. You should be able to get the engine running again by pressing on the flexible metal tab on the stop switch and reattaching the wire. Take care not to break the wire as you twist it back into position. Also, check to make sure that the spark plug lead is properly attached to the plug.

How to Degrease an Engine

1

2

3

1. With your equipment in a well-ventilated area and the engine off, spray a degreasing agent, such as those available at auto parts stores, liberally on greasy and dirt-encrusted surfaces. Wait fifteen minutes to allow grease and dirt to break down.

2. Wipe away the residue with a clean cloth.

3. Hose off the equipment surfaces and then allow them to dry completely before storing.

Fuel tanks must be constructed of a noncorrosive material or coated with a corrosion-resistant layer to protect against the damaging effects of water, alcohol, and salt. If the tank is designed to deliver fuel through a fuel line, a convex fuel filter may be located at the base of the tank, where fuel from the tank enters the fuel line. A filter can also be located outside the tank, midway along the fuel line. Many older engines were not equipped with a fuel filter, though a generic one may be added if there's sufficient room to cut the fuel line (typically at center length) and insert it.

Servicing the Fuel Tank

Fuel tanks are designed to keep your engine's fuel clean, vented, and secure. If you spot debris in the tank or leaking gasoline, it's time for fuel tank maintenance. If you find a crack or a hole, a new tank is the answer. Older models are made of steel. Newer models are made of aluminum or plastic. In either case, don't repair a damaged tank. It poses a risk of leakage, fire, or fuel contamination.

Tanks are typically installed as far away as possible from the hottest areas of the engine to keep the fuel cool. Other factors can also damage your tank. It can crack due to long-term exposure to hot sun and other elements, or if it is used to support weight from outside. Tanks often crack under such stress. Sometimes the tank hose flange gets cracked or broken, too. If you're replacing a fuel tank, use only parts recommended by the engine manufacturer. These parts will attach securely to your engine in the space provided. Many fuel tanks

are designed to use a vented fuel cap to prevent a vacuum from forming in the fuel line. Check to determine if the vent hole is obstructed. If fuel is leaking from the cap, a properly fitted replacement cap can solve the problem.

NOTE: Your tank may also contain a fuel filter (see "Servicing the Fuel Filter," pages 75 to 76). Check it occasionally for debris and signs of water.

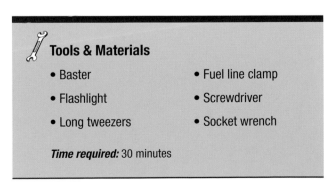

Tools & Materials

- Baster
- Flashlight
- Long tweezers
- Fuel line clamp
- Screwdriver
- Socket wrench

Time required: 30 minutes

LABYRINTH FUEL TANKS

If your tank must sustain excessive vibrations, you can install a labyrinth-equipped tank on some engine models. The labyrinth, available from your authorized service dealer, contains a set of baffles and/or a foam insert to reduce the sloshing and vaporization of fuel.

How to Remove and Clean the Fuel Tank

1. Remove the spark plug lead and secure it away from the plug. Use a fuel line clamp or other smooth-faced clamp (such as a 1- to 3-inch C-clamp) to seal the fuel line where it attaches to the carburetor. Then, disconnect the line from the carburetor, hold the line over a bucket or fuel can and release the clamp. Dispose of all fuel in a safe manner (see "Gasoline Use," page 77).

2. Check with a flashlight for debris and beads of light that indicate holes or cracks. Use a baster or long tweezers (to grasp solids) to remove loose debris from the tank. Once the tank is free of liquids, it can be shaken upside down with the cap off in order to evacuate stubborn debris. If you find damage that questions the tank's integrity, replace the tank with original manufacturer's equipment. Inspect the fuel filter for debris or deposits (see next page). Reattach the fuel tank or install a new tank, fastening it firmly with the cap screws. This is a good opportunity to replace the fuel line and filter, using original manufacturer's equipment or quality generics recommended for your machine.

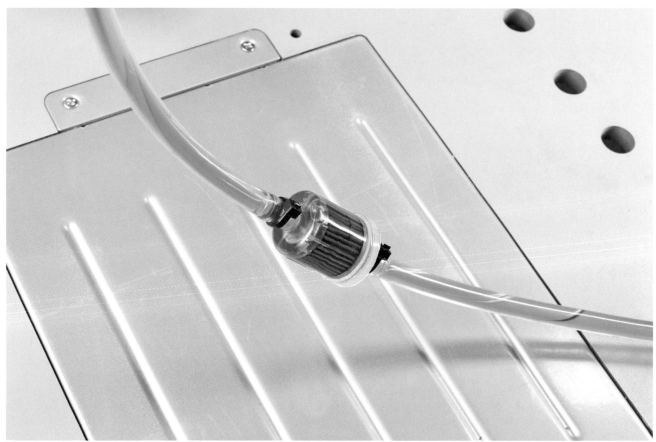

On rare occasions the fuel tank may contain a fuel filter, but in almost all cases it will be found in-line on the fuel line leading from the tank.

Servicing the Fuel Filter

A clean fuel filter strains the fuel before it reaches the carburetor and prevents foreign particles from clogging your engine. A dirty fuel filter can make the engine run too lean, with diminished performance and uneven operation. Other factors can cause these problems, but the fuel supply is one of the easiest to check.

Some filters are located inside the tank, others are fitted into the fuel line between the tank and the fuel pump. Most use either a mesh screen or pleated paper. The size of the holes in the filter will determine the largest particles that can get through the filter, and the number of holes will affect the amount of fuel that can flow through the filter.

Filters contain either a mesh screen or a pleated-paper element, and are rated by the size of the holes in the filtering material, expressed in microns (μ). Pleated-paper filters, designed for use in the fuel tank, are typically contained in a clear plastic casing and rated 60μ. They consist of multiple folds that strain out particles suspended in the fuel.

The proper filter for your engine depends on the engine's design. Consult your owner's manual or small-engine parts retailer for the correct replacement filter.

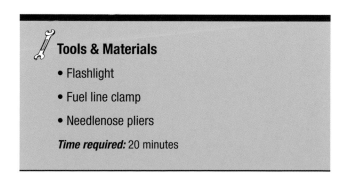

Tools & Materials

• Flashlight

• Fuel line clamp

• Needlenose pliers

Time required: 20 minutes

How to Inspect a Fuel Filter

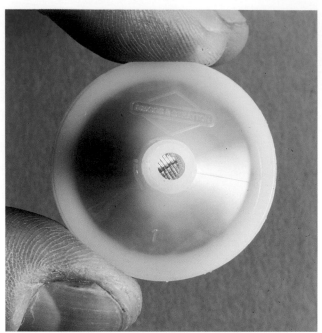

1. Shut the fuel valve, if the engine is equipped with one. It's located at the base of the fuel tank, where the fuel line is attached. If your tank is not equipped with a fuel valve, clamp the fuel line, using a fuel line clamp (see page 74). If your filter is installed in the fuel line, remove the metal clips on each side of the filter, using needlenose pliers, and slide the filter out of the fuel line.

2. Shake the filter over a clean cloth to displace any remaining fuel, then use the cloth to wipe away any residue from the outside of the filter. Keep the filter a safe distance from your face and look into one end. You should be able to see light shining through clearly from the other side. If anything is clogging the mesh screen or pleated paper or the inside of the casing, replace the filter.

SAFETY TIP

Wear safety eyewear whenever removing or inspecting a filter to protect your eyes from liquid fuel or fuel vapors. Have a dry cloth handy to hold the filter and catch any dripping fuel. If the filter is installed inside the tank, you will need to drain the tank before you can remove the filter for inspection or replacement.

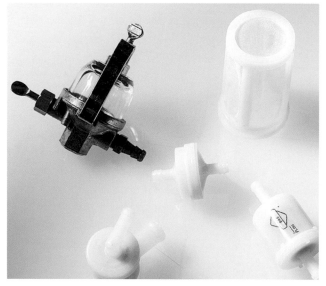

Heavier-duty small engines contain a fuel filter. A sampling of in-line filters are seen above. If you don't find a filter mounted in your fuel line, look for one inside the fuel tank (see page 75).

WHICH GASOLINE TO USE?

Use only fresh unleaded gasoline in your small engine. Here are a few other tips for gasoline use in 4-stroke small engines.

- Use gasoline with a 77 octane rating or higher for L-head engines and 85 or higher for overhead valve engines. Since small engines operate at relatively low compression ratios, knocking is seldom a problem, and using gasoline with a higher octane rating is unlikely to offer any benefit.

- Using gas that is over a month old and does not contain a gasoline stabilizer may result in hard starting and varnish formation. Drain the tank if fuel sits for more than a month. The Environmental Protection Agency recommends pouring old fuel into your car's gas tank. As long as the car's tank is at least half full, the old fuel will mix harmlessly with the new and will not affect your car engine.

- Four-stroke engines are used on most lawn mowers and large lawn equipment. NEVER use an oil-gasoline mixture in a 4-stroke, since the engine has an independent oil supply. If you find an oil fill cap leading to the crankcase, you can be sure your engine is a 4-stroke.

BEWARE OF ETHANOL

Ethanol blended gasolines are no friends to small engines, especially the older ones! Its alcohol content tends to eat rubber gaskets, draw moisture out of the ambient air (think *damp garage* and a mower or snowblower that sits around breathing moist air for weeks!), and promotes corrosion.

Even new engines built around the Ethanol reality can be subject to the negative impacts of this 10 percent additive in motor fuel. And, increased Ethanol percentages, such as in E-85 or other so-called "flex-fuels" can really cause a small engine big problems, so should *never* be fed to your little motor built as of this book's publication date.

Though not needing the higher octane of an unleaded and Ethanol-free premium fuel available from gas stations offering an alternative to Ethanol, some lucky small engines are fed this fuel by their owners who avoid being "penny wise and pound foolish." Others treat their small engine's fuel with "Ethanol additive cures" available at auto parts stores.

Admittedly, gasoline that contains Ethanol is a major subject of controversy in small-engine/outboard motor circles. No doubt, anyone who mentions the topic will get ample commentary. If your small-engine is new enough that you know where it came from, consulting the dealer is a good avenue of specifically helpful advice.

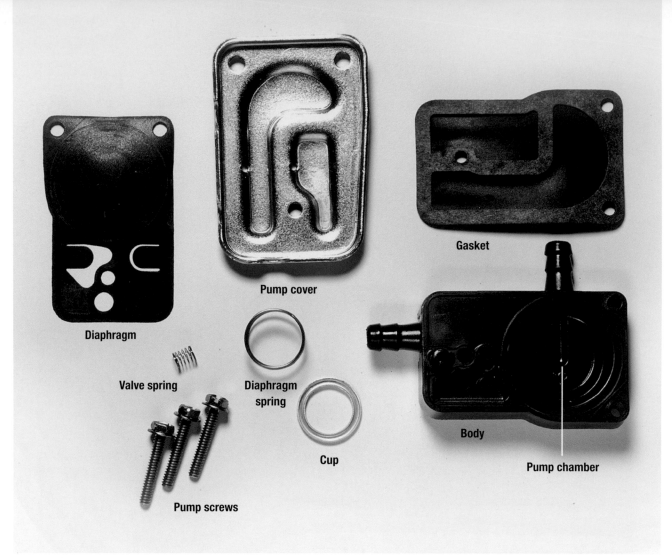

Diaphragm

Pump cover

Gasket

Valve spring

Diaphragm spring

Body

Cup

Pump chamber

Pump screws

The parts of a typical small-engine fuel pump. The device relies upon vacuum pressure generated by the piston motion to advance the fuel upline from the tank.

Servicing the Fuel Pump

A fuel pump is used when the fuel tank is mounted lower than the carburetor and cannot rely on gravity to carry fuel through the fuel line. Fuel pumps have either a plastic or a metal body and develop pressure using the vacuum in the crankcase, which is created by the motion of the piston. A fitting on the crankcase cover or the dipstick tube draws on the crankcase vacuum to create the pressure to pump fuel.

The fuel pump may be mounted on the carburetor, near the fuel tank or between the tank and carburetor.

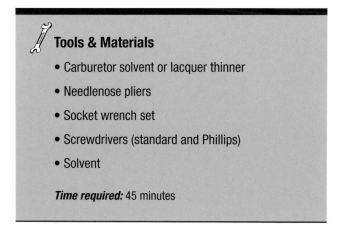

Tools & Materials

- Carburetor solvent or lacquer thinner
- Needlenose pliers
- Socket wrench set
- Screwdrivers (standard and Phillips)
- Solvent

Time required: 45 minutes

How to Inspect a Fuel Pump

Metal-body pump

1. Turn off the fuel valve (if equipped) at the base of the fuel tank, where the fuel line is attached. If there is no fuel valve, stop the flow of fuel, using a fuel line clamp. Loosen the mounting screws and remove the pump from the mounting bracket or carburetor.

Hairline crack

2. Check for hairline cracks and other damage to the external surfaces of the pump. If the pump is damaged and has a metal body, discard the pump and install a replacement pump from the engine manufacturer.

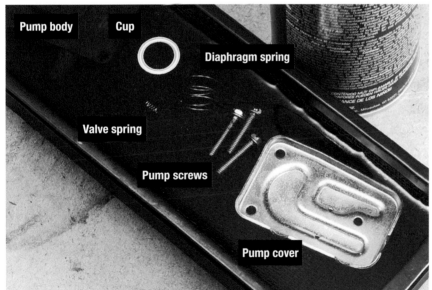

Pump body   Cup

Diaphragm spring

Valve spring

Pump screws

Pump cover

PLASTIC BODY OPTION 1: If the pump has a plastic body, obtain a manufacturer's repair kit so you can replace worn parts. Check the hoses for cracks, softening or hardening, and replace any faulty parts. Discard old gaskets, diaphragms, and springs, and replace them with new parts from the kit.

PLASTIC BODY OPTION 2: With the fuel valve closed or the line clamped, remove the mounting screws. Then, disconnect the fuel hoses, using needlenose pliers to loosen the clips. Remove the screws and disassemble the pump. Inspect the body for cracks or other damage. Soak metal parts in all-purpose parts cleaner. The pump body may be soaked for up to fifteen minutes.

REINSTALLING THE PUMP

Here is how you reinstall a fuel pump: Referring to the photo on page 78, place the diaphragm spring and then the cup over the center of the pump chamber. Also insert a valve spring. Install the diaphragm, gasket and cover and attach with pump screws. Tighten the screws to 10 to 15 inch-pounds, using a torque wrench. Attach the pump to the carburetor or mounting bracket, using the pump mounting screws.

The small carburetors found on most small engines perform essentially the same function as an automotive carburetor (prior to fuel injection). They contain the choke plate that allows fuel and air into the combustion chamber. Also like their automotive counterparts, they occasionally need to be cleaned, adjusted, rebuilt, or replaced.

Servicing the Carburetor

A big part of ensuring a smooth-running engine is keeping your carburetor and linkages clean and well adjusted. The linkages attached to the carburetor's throttle and choke plates can bind or stick when dirty. Constant vibration and wear can affect the settings of the carburetor's mixture screws.

With all of the grass, twigs, and other debris that a small engine encounters, it's not surprising that even passages inside the carburetor eventually pay a price. Deposits inside the carburetor can clog fuel and air passages and reduce performance or stop the engine altogether. Luckily, you can take care of many of these problems quickly and easily—often without even removing the carburetor from the engine.

Tools & Materials

- Carburetor cleaner (spray-type)
- Safety eye wear
- Screwdrivers (Phillips and standard)
- Socket wrench set
- Tachometer

Time required: 1 hour

How To Find the Source of a Fuel Supply Problem

Remove the air cleaner and inspect the choke plate mounted on a shaft at the opening of the carburetor's throat. Check that the choke plate closes easily and completely. A choke that does not move freely or close properly can cause difficulties in starting. Spray a small amount of carburetor cleaner on the shaft of a sluggish choke and into the venturi to loosen grit. Dab the parts dry with a clean rag. Debris in the carburetor often causes performance problems.

Open the fuel valve (if equipped), located at the base of the fuel tank where the fuel line is attached. Remove the line and check for blockage. Fuel will not reach the carburetor if the fuel valve is closed. Also, if the engine is equipped with a fuel pump, make sure it operates properly (see "Servicing the Fuel Pump," pages 78 to 79).

Remove and inspect the spark plug. A wet plug may indicate over-choking, water in the fuel (see "Servicing the Fuel Tank," pages 78 to 79), or an excessively rich fuel mixture. A dry plug may indicate a plugged fuel filter (see "Servicing the Fuel Filter," pages 75 to 76), leaking mounting gaskets on either end of the carburetor, or a stuck or clogged carburetor inlet needle.

Inject a small amount of combustible fluid into the spark plug hole: to avoid spillage, try spraying a little bit of starting fluid into the hole through the nozzle extension or with an automotive funnel. You can also use a dab of gasoline, but you should avoid handling fuel as much as possible. Once the combustible fluid is in the hole, screw the spark plug back in and start the engine. If it fires only a few times and then quits, assume a dry plug condition and consider the causes of a dry plug, listed in the previous step.

How to Adjust the Idle Speed & Mixture

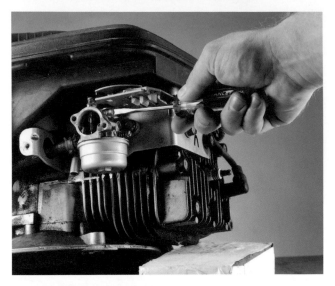

1. On some float-type carburetors, you can adjust the air-fuel mixture and engine speed at idle. Check for an idle speed screw designed to keep the throttle plate from closing completely (see "Parts of the Carburetor," page 101) and an idle mixture screw that limits the flow of fuel at idle. If your carburetor contains these screws, remove the air filter and air cartridge. Locate the idle mixture screw and turn it clockwise until the needle lightly touches the seat. Then, turn the screw counterclockwise 1½ turns. If your carburetor has a main jet adjustment screw at the base of the float bowl, turn the screw clockwise until you feel it just touch the seat inside the emulsion tube. Then, turn the screw counterclockwise 1 to 1½ turns. Replace the air cleaner assembly and start the engine for final carburetor adjustments. Run the engine for five minutes at half throttle to bring it to its operating temperature. Then, turn the idle mixture screw slowly clockwise until the engine begins to slow. Turn the screw in the opposite direction until the engine again begins to slow. Finally, turn the screw back to the midpoint.

2. Using a tachometer to gauge engine speed, set the idle speed screw to bring the engine to 1750 rpm for aluminum-cylinder engines or 1200 rpm for engines with a cast-iron cylinder sleeve.

3. With the engine running at idle (inset photo), hold the throttle lever against the idle speed screw to bring the engine speed to "true idle." Then, repeat the idle mixture screw adjustments from Step 1 to fine-tune the mixture.

ADJUSTING THE HIGH-SPEED MIXTURE

Some older carburetors contain a high-speed mixture screw, near the throttle plate and opposite the idle speed screw. Under load, the high-speed circuit increases air flow through the throat. Setting the high-speed mixture involves running the engine until it is warm, stopping it to adjust the high-speed mixture and then restarting for final adjustments. Run the engine for five minutes at half throttle to bring it to its operating temperature. Then, stop the engine. Locate the high-speed mixture screw and turn it clockwise until the needle just touches the seat. Then, turn the screw counterclockwise 1¼ to 1½ turns. Restart the engine and set the throttle position to HIGH or FAST. Turn the high-speed or main jet screw clockwise until the engine begins to slow. Then, turn the screw the other way until the engine begins to slow. Turn the screw back to the midpoint.

Once adjusted, check engine acceleration by moving the throttle from idle to fast. The engine should accelerate smoothly. If necessary, readjust the mixture screws.

ADJUSTING THE CHOKE LINKAGE

Remove the air cleaner and locate the choke lever on the engine or on the remote engine speed controls. Move the equipment controls to FAST or HIGH (left). Loosen the cable mounting bracket to allow movement of the cable casing. Move the cable casing so the choke is closed. Tighten the cable mounting screw (right photo) and check the motion of the control lever. Repeat the steps, as necessary, until the cable moves freely.

Adjusting the Governor

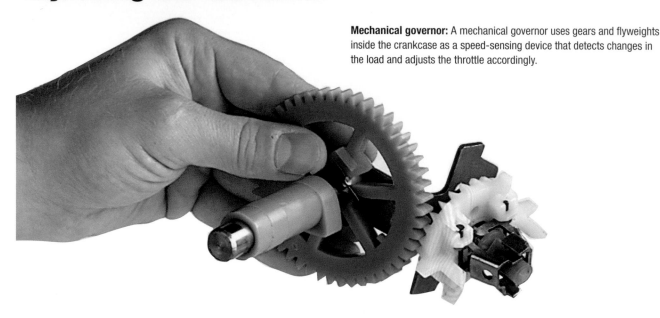

Mechanical governor: A mechanical governor uses gears and flyweights inside the crankcase as a speed-sensing device that detects changes in the load and adjusts the throttle accordingly.

A properly adjusted governor can maintain a steady engine speed regardless of changes in the terrain and other conditions that increase the work of the engine. These conditions are known as the "load." When engine speed starts to rise or fall in response to a change in the load, the governor responds, opening or closing the throttle. If you adjust engine speed manually, using the equipment controls, the governor's job is to maintain the new setting.

Your engine contains either a pneumatic governor or a mechanical governor (see "Governor System," page 25). Remove the blower housing to determine which one your engine uses. Pneumatic governor linkages connect to a pivoting air vane next to the flywheel. On a mechanical governor, the linkages connect to a governor shaft (see photo, above).

For either type, follow the steps in these pages to adjust your governor for best performance. NOTE: Governor adjustment procedures vary widely depending on the make and model of the engine. Check with your authorized service dealer for the speed settings for your equipment.

The tang bending tool is the most common tool for setting governor spring tension. It's a simple metal lever with forked ends for bending the tabs, or tangs, on the governor and other engine parts.

Tools & Materials

- Standard screwdriver
- Combination wrench set
- Socket wrench set
- Tang bending tool

Time required: 45 minutes

PARTS OF THE MECHANICAL GOVERNOR

1. Engine speed control. Moving this lever to a higher speed setting opens the throttle indirectly by pulling on the governor gear bracket.

2. Governor gear bracket. The bracket pivots, increasing tension on the governor spring.

3. Governor spring. Tension on the spring pulls on the governor lever in an effort to open the throttle plate.

4. Governor lever. The lever pivots, pulling on the throttle linkage and applying pressure to the governor shaft.

Crankshaft

5. Governor shaft. The shaft links the governor linkages and levers to the governor cup and other parts inside the crankcase.

6. Throttle linkage. The linkage tugs on the throttle lever.

7. Throttle lever. The lever opens the throttle plate, allowing more air-fuel mixture into the combustion chamber, causing engine speed to increase.

8. Governor gear. Increased engine speed causes governor gear to spin faster and flyweights to fly outward.

9. Flyweights. Movement of the flyweights applies pressure to the governor cup.

10. Governor cup. Governor cup causes the governor lever to pivot.

Spark plug lead

Choke control

INSPECTING THE GOVERNOR

1. With the spark plug lead disconnected and secured away from the spark plug, check that the governor linkages are attached and move freely by pulling gently on the throttle lever. This should stretch the governor spring while pressing on the governor lever. If not, check that the governor spring and the link to the governor lever are properly attached to the throttle lever.

2. Springs and linkages that are not attached may be reconnected if they are in good condition. Twist them carefully into place to ensure that the delicate springs and linkages aren't permanently bent or stretched. Do not use pliers or other tools to bend or distort links or springs. Replace the governor spring if it is overstretched and replace the linkages if they appear worn.

HUNTING AND SURGING

Your engine may race or slow intermittently even when the load and the speed control settings are unchanged. Follow the steps below to determine whether the source of this erratic engine behavior—known as "hunting and surging" is the carburetor or the governor.

1. Check that springs and linkages move freely and that the governor spring is inserted properly on the governor lever arm (above). On a mechanical governor, perform a static governor adjustment (opposite).

2. Run your engine at each of its speed settings to determine when hunting and surging occur. If the problem crops up at "true idle" (when the throttle lever is against the idle speed screw or stop), the air-fuel mixture is the likely cause. An air leak or debris in the carburetor is probably causing the air-fuel mixture to fluctuate. Remove and clean your carburetor (see pages 100 to 107).

3. If hunting and surging occur at top no-load speed, run the following test. Move the throttle lever so the throttle plate is in the open

position. If hunting and surging continue, the problem is probably in the carburetor. Clean and adjust the carburetor (see "Overhauling the Carburetor," page 100). If hunting and surging is eliminated by opening the throttle, lubricate the governor linkage to eliminate any resistance and binding. If hunting and surging persist, replace the governor spring(s) and retest.

4. Some engines have a separate "governed idle" spring and governed idle adjusting screw to prevent stalls under light loads.

If hunting and surging occurs under light load, run the following test. Look for an idle speed screw, a stop screw on top of the carburetor, designed to prevent the throttle from closing completely. Hold the throttle lever against the idle speed screw and increase the governed idle speed by turning the screw slowly clockwise. If hunting and surging stop, replace the idle spring and linkage and reset the governed idle speed. If hunting and surging continue, clean and adjust the carburetor (see "Overhauling the Carburetor," pages 100 to 107).

ADJUSTING THE "STATIC" SETTING ON A MECHANICAL GOVERNOR

The following procedure eliminates play in a mechanical governor between the governor crank—the arm that protrudes from the crankcase—and governor system components inside the crankcase. This procedure does not apply if your engine has a pneumatic governor.

1. Loosen the clamp bolt on the governor crank until the governor lever moves freely.

2. Move the throttle plate linkage until the throttle plate is wide open. (To find the wide-open position, first position the throttle lever against the idle speed screw or a fixed stop plate. The throttle is wide open when it is all the way in the opposite direction.) Note the governor arm's direction of rotation as you move the throttle plate to the wide-open position. This is important for the next step.

3. With the throttle plate wide open, use a nut driver or wrench to turn the governor shaft in the same direction that the governor arm traveled.

4. Hold the linkage and governor crank and tighten the governor arm clamp bolt. Move the linkage manually to make sure there is no binding.

ADJUSTING GOVERNED IDLE

Some engines contain a shorter, smaller, "secondary" governor spring to discourage stalls when the engine is operating at idle under a light load. Under these conditions, the secondary spring keeps the engine at a "governed idle" speed slightly above its true idle speed. The idle speed screw is always set at less than the engine's governed idle speed. The procedure for adjusting governed idle varies depending on the engine model. Consult your owner's manual for the procedure for your model.

Keep in mind that the secondary spring affects all governor settings. If the governor on your engine has a secondary spring, you need to adjust the governed idle before setting the engine's top no-load speed.

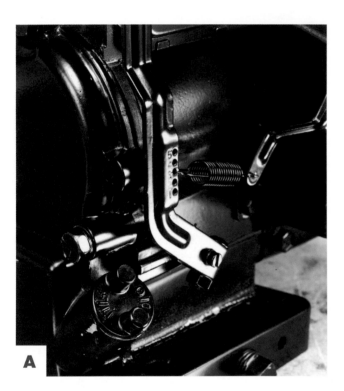

A

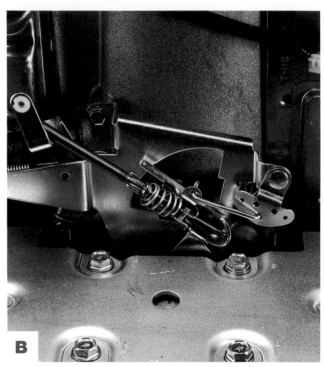

B

TANG BENDING AND OTHER ADJUSTMENT METHODS

The tang bending tool pictured on page 84 (bottom) is the most common tool for governor spring adjustment. It's a simple metal lever with forked ends used for grasping and bending the tabs, or tangs, on the governor lever, spring anchor, and other engine parts. Bending a tang increases or decreases the extension of the governor springs.

If the governor lever has multiple spring holes (photo A), you can increase top no-load speed by selecting a hole that is farther from the pivot point on the governor lever. On some engines, an adjustment screw alters governor spring tension, increasing or decreasing top no-load speed (photo B). Fine adjustments may still require the use of a tang bending tool (photos D, E, and F).

SETTING TOP NO-LOAD SPEED

If your engine races when you set your controls to HIGH, you need to reduce the engine's top speed under no-load conditions. Ask your authorized service dealer for the proper top no-load speed setting for your model. (If your governor contains two springs, skip to "Setting Dual-Spring Top No-Load Speed," below). The most common method for adjusting top no-load speed is to use a tang bending tool (see "Tang Bending and Other Adjustment Methods," opposite page) to bend the spring anchor tang to stretch or relax the spring.

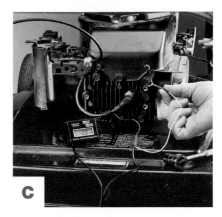

1. Attach the Tiny Tach to the engine's white ground wire with the alligator clip. The red wire should be wrapped around the spark plug lead (photo C). Run the engine for five minutes so it reaches its operating temperature.
2. Place the equipment on a hard, smooth surface with the engine running and the controls set to HIGH. Decrease top no-load speed by bending the tang toward the governor spring, until the manufacturer's specified speed setting is attained (photo D). Increase top no-load speed by lengthening the spring.
3. If your engine has a mechanical governor, proceed with the static governor adjustment (see "Adjusting the 'Static' Setting on a Mechanical Governor," page 87).

SETTING DUAL-SPRING TOP NO-LOAD SPEED

If your governor contains two springs, the smaller, shorter spring is the secondary spring and must be adjusted to prevent stalls.

1. Attach a tachometer to the engine. With the engine running, bend the secondary spring tang (see "Tang Bending and Other Adjustment Methods," opposite page) so there is no tension on the secondary spring (photo E).
2. Bend the primary governor spring tang until the engine speed is 200 rpm below the manufacturer's specified top no-load speed (photo F).
3. Bend the secondary governor spring tang until the engine reaches its top no-load speed. Ask your authorized service dealer for the top no-load speed setting for your engine.
4. If your engine has a mechanical governor, proceed with the static governor adjustment (see "Adjusting the 'Static' Setting on a Mechanical Governor," page 87).

Replacing the Rewind

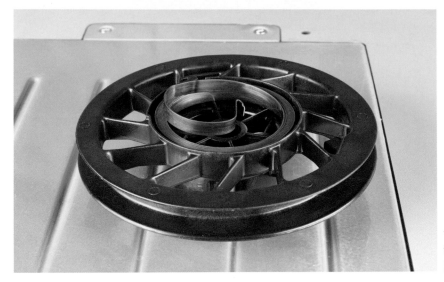

Tools & Materials

- Needlenose pliers
- Power drill
- Socket wrench set

Time required: 45 minutes

The rewind assembly consists of a pull rope, a flywheel, and a rewind spring that causes the pull rope to recoil.

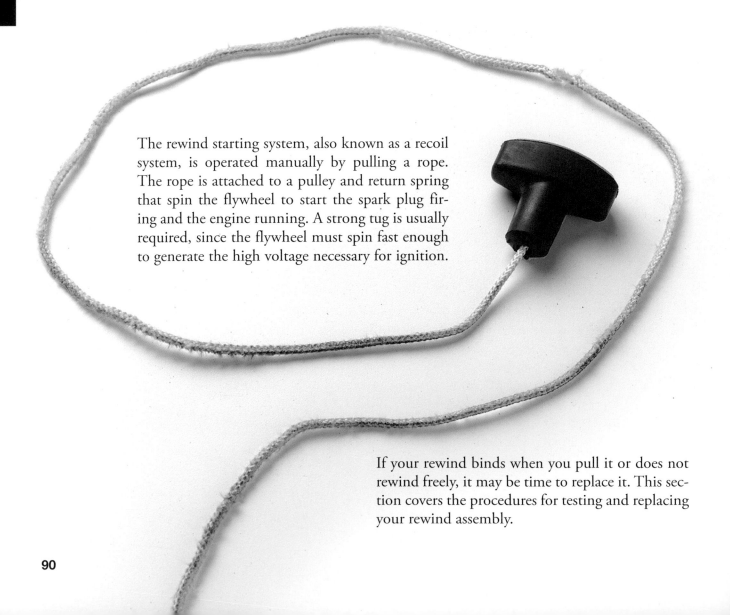

The rewind starting system, also known as a recoil system, is operated manually by pulling a rope. The rope is attached to a pulley and return spring that spin the flywheel to start the spark plug firing and the engine running. A strong tug is usually required, since the flywheel must spin fast enough to generate the high voltage necessary for ignition.

If your rewind binds when you pull it or does not rewind freely, it may be time to replace it. This section covers the procedures for testing and replacing your rewind assembly.

TESTING THE REWIND

If you discover any of the following conditions, replace the entire rewind assembly.

1. Pull the rope slowly (photo A). If the system is noisy, binds, or feels rough, the return spring, pulley, or rope may be jammed. If the crankshaft doesn't turn, the ratcheting mechanism isn't engaging.
2. Once the rope is all the way out, check for fraying and wear along its entire length.
3. Let the rope rewind slowly. If it fails to rewind, the pulley may be binding or the return spring may be broken, disengaged, or worn.

REPLACING THE REWIND

On some engines, the rewind is spot-welded or riveted to the top of the engine shroud. On others, it is attached with nuts or bolts.

1. Loosen the appropriate bolts and remove the blower housing (see "Removing Debris," pages 69 to 71).
2. Remove the nuts or bolts on the rewind (if equipped) or drill out the rivets or spot welds with a ³⁄₁₆-inch bit, drilling only far enough to loosen them (photo B).
3. Install a replacement rewind from the original engine manufacturer. Insert the mounting bolts from inside the blower housing so that the bottoms come through the top of the shroud. Place the replacement rewind over the bolts and fasten a washer and nut securely on each bolt (photo C).

REWIND SAFETY

A rewind assembly contains a pulley and spring that retract the rope after each pull. Disassembling a rewind is best left to a small-engine technician. The project requires special care and safety precautions because of the risk of serious injury from a spring or other flying parts.

A

Old rewind

B

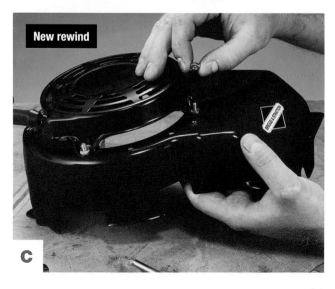

New rewind

C

Inspecting & Changing the Muffler

Tools & Materials

- Hammer
- Mallet
- Metal snips
- Pipe wrench
- Pin punch
- Slip-joint pliers
- Socket wrench

Time required: 30 minutes

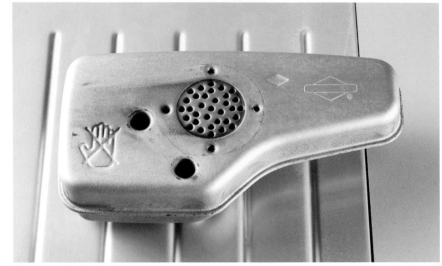

When your small engine's muffler is ready for replacement you will know just from the engine noise. These important parts are not repairable: better to find the correct replacement muffler and swap it in.

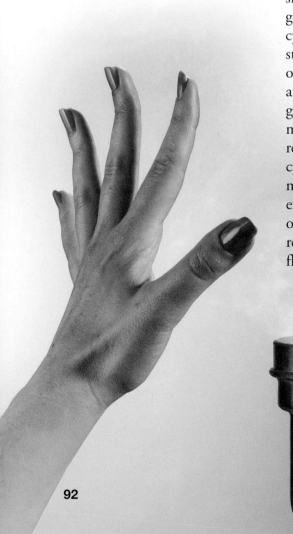

One of the main sources of small-engine noise is the hot gases that are forced out of the cylinder during each exhaust stroke. A muffler does a good job of reducing exhaust noise. But after a season or two, exhaust gases leave a layer of soot in the muffler that creates additional resistance to gases exiting the cylinder. When hard soot accumulates, when the exhaust emits excessive noise, or when cracks or holes appear, don't try to repair the muffler. Once a muffler shows signs of deterioration, replace it. This section covers the procedures for removing, inspecting and replacing your muffler. It's an inexpensive and simple job if you take the proper precautions. Always wait for the engine to cool completely before handling the muffler. The muffler's surface can remain very hot and can easily cause a burn, even long after the engine is stopped. A rusty muffler can cut you, especially if it crumbles during replacement. If any sharp edges are exposed, use slip-joint pliers to remove the muffler.

TIP

HOT! Keep your distance from the muffler and other engine parts until the engine has had plenty of time to cool.

HOW YOUR MUFFLER WORKS

The force of exhaust gases as they rush through the small opening in the exhaust valve produces shock waves. It's the muffler's job to reduce noise by routing the exhaust through a series of perforated baffles and plates that break up the sound waves. The inside of the muffler also functions as a spark arrestor, preventing exhaust sparks from exiting and igniting dry grass, leaves, or debris.

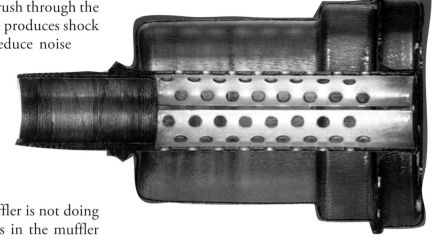

We quickly notice when a muffler is not doing its job. Even small cracks or holes in the muffler can result in a dramatic increase in engine noise.

When it's time to replace the muffler, use original manufacturer's equipment or a quality generic suitable for your model, thus promoting safety and optimal engine performance. As you can see from this photo, mufflers come in a very wide array of sizes and shapes.

INSPECTING THE MUFFLER

1. Locate the muffler, which is usually near the cylinder head.

2. Check the outside of the muffler for signs of rust, dents, holes, or cracks, any of which can restrict the exhaust and reduce the effectiveness of the muffler.

3. If there's soot near the exit hole, remove the muffler to check for soot inside.

REMOVING THE MUFFLER

The muffler body may be attached directly to the engine with mounting bolts or screwed into the engine body. On some mufflers, an extended pipe threads into the engine.

1. If the muffler is attached with mounting bolts and has locking tabs around the bolts, bend the tabs back far enough to fit a wrench over the bolt heads. Remove the bolts (photo A) and detach the muffler.

A

   If the muffler screws directly into the engine, apply some penetrating oil/lubricant to the threads (photo B) and let the oil work for several minutes. Tip the engine very slightly, if necessary, to allow the oil to reach the threads. NOTE: Don't tip the engine sharply. On a 4-stroke, this can cause oil to drain into the carburetor and air cleaner.

B

Some mufflers are fastened with a threaded lock ring. Loosen it by tapping it counterclockwise with a hammer and pin punch. Then, grasp the muffler with slip-joint pliers and unscrew counterclockwise (photo C).

2. To check for soot, tap the muffler body with a mallet or on a hard surface (photo D). If the muffler is damaged or large quantities of soot cannot be dislodged, replace the muffler with original manufacturer's parts. If the muffler is in good condition, reattach it.

3. Don't overtighten the new muffler. If a lock ring is used, install it using a hammer and pin punch. NOTE: The smooth side of the lock ring must be against the cylinder in aluminum-block engines. The tooth side must be against the cylinder in cast-iron engines.

4. Brush the entire area to clear away dirt and debris. If left on the muffler, dried grass clippings and other debris can catch fire on the hot surface of the muffler.

C

D

REMOVING A RUSTY MUFFLER

A very rusty muffler may collapse or crumble as you twist it with a wrench. There's no harm done as long as you take care not to damage muffler mounting threads in the engine block or other muffler fittings on the engine. If your muffler screws into the engine, cut the muffler body off with metal snips. Then, grasp the stem using slip-joint pliers, and unscrew (photo E). If the muffler breaks off, leaving a connecting pipe attached to the engine, grasp the pipe with slip-joint pliers and unscrew (photo F).

E

F

Advanced Repairs

The systems and techniques in this section are more complex than those described in the Basic Repair section, but they're well within your reach now that you have some primary repair projects in your repertoire, as the same basic principles apply. If an advanced repair task is new to you, start by reviewing "Trouble-shooting" (OLD pages 60 to 61) and "Safety" (OLD pages 12 to 13).

Here are a few of the subjects you should review to prepare for the projects in this section:

- Before you clean your carburetor, review the basics of carburetor operation (see "Fuel System," pages 38 to 41).
- Before you replace the ignition, you should have some knowledge of how electricity is generated (see "Ignition System," pages 20 to 21).
- Before removing carbon deposits, it's helpful to understand the source of carbon deposits in the engine (see "Compression System" and "Fuel System," pages 12 to 19, and "Lubrication & Cooling System," pages 22 to 24).
- Before servicing the valves in a 4-stroke engine, review the mechanics of compression (see "Compression System," pages 12 to 16).
- Before servicing a brake or stop switches, review the principles of the flywheel brake (see "Braking System," page 26).

The background provided in these system sections will help you get through the advanced repairs projects with no trouble. If you're concerned about special issues that pertain to your engine, contact an authorized service dealer. The technicians there can offer you helpful advice specific to your engine make and model.

If you have experience in basic repair and confidence in your ability, you can take on these tasks by yourself:

Inspecting the Flywheel & Key

Overhauling the Carburetor

Testing the Electrical System

Replacing the Ignition

Servicing the Valves

Removing Carbon Deposits

Servicing the Brake

THE ABANDONED TRASH-DAY LAWN MOWER

It's amazing what some people will throw out! For outdoor power equipment buffs on a budget or with a penchant for revitalization, mowers wheeled to the curb to await their swan-song ride to the dump represent a great opportunity to exercise your skills in advanced repair.

Upscale neighborhoods often hold the biggest selection of prematurely disposed of lawn mowers. But there are folks at all kinds of addresses who get something new and consider selling their old gear to be too much hassle. During a few weeks recently, I noted five such rejected machines within a mile of my home. Admittedly, two were true goners, but one—with 2-stroke power and most of its original light green paint still gleaming—looked so good that I had to consult its owner as to whether the unit was truly bound for the recycler. "Take it!" he welcomed with an enthusiastic gesture. The fellow then pointed to its 4-stroke recent replacement and explained that he was simply tired of having to mix gas and oil for his former mower.

Another imminent inmate of the landfill got quickly rescued when a passerby ascertained its only problem was a faulty drive mechanism. "Wow! The engine and deck look great!" the delighted finder exclaimed, "Even if I can't get the wheel-drive to engage," he noted while heading away with his treasure, "I can always use it as a push mower."

Because time is of the essence when spotting a mower amongst the garbage cans, the following list is designed for making a quick thumbs-up or down:

- *Engage the rewind starter.* Does the piston move freely in the cylinder? Is there sufficient compression or does a sharp pull feel weak? Hear any clunks or roughness? Might the engine brake still be engaged even with the bail in the operating position? If so, see if it's just a cable issue. If it pulls hard or not at all (and the issue can't be attributed to a stuck rewind) walk away.

- *Check the oil on a 4-stroke.* Is there any? Is it dirty/pitch black? Some is better than none, but presence, color (the lighter/cleaner the better), and the hopeful absence of water are valid indicators of past care.

- *Check the fuel.* Water in the fuel isn't a good sign of the mower's history, but its typically not a "deal-breaker" on a free (or bargain-priced garage sale) mower. Get rid of the H_2O as soon as you get it home.

- *Look carefully at the deck.* Is there any fatal rust-through (especially where the wheels, engine, and push handle connect to the deck)? Are all of the wheels intact?

- *Check under the deck.* Obviously bent and severely nicked blades serve as tell-tales to the machine's use. The cleaner the underside, the better owner care the mower enjoyed. If the area under the engine is oily (unless the muffler outputs there), there's probably a seal/bearing issue. Stay clear of that mower! The same assessment applies for any signs of a bent crankshaft (on which the blade is attached). Check for this major negative by first removing the spark plug wire and then tipping the mower so the crankshaft travel can be noted by carefully rotating the blade.

- *Check for spark.* If all else is OK, remove the spark plug, reattach the spark plug lead/wire, and with the base of it (where the threads are) touching the engine block, engage the starter while looking for a spark at the electrode in the bottom of the plug. Some small-engine hobbyists figure the ignition system will need attention anyway, so don't bother with this test on an otherwise "clean" motor until they get their find home. (Note: Much of this test regimen applies to other forlorn outdoor power equipment left for the sanitation department crew.)

—P. H.

Tools & Materials

- Box wrenches
- Carburetor cleaner
- Carburetor repair kit
- Compressed air
- Enamel nail polish
- Fuel line clamp
- Hammer
- ⁵⁄₃₂-inch pin punch
- Screwdrivers (standard, carburetor-type)
- Torque wrench

Time required: 90 minutes

The parts found in a typical carburetor rebuild kit.

Overhauling Carburetors

Most carburetor problems are caused by dirt particles, varnish, and other deposits that block the narrow fuel and air passages inside. Gaskets and O-rings are also common sources of problems. They eventually shrink, causing fuel and air leaks that lead to poor engine performance.

If you're doing a rebuild, you'll need to purchase the repair kit for your carburetor, which includes replacement gaskets and other necessary parts. While you're at it, check the price of a complete replacement carburetor for your engine. In some cases, it may be more cost-effective to install a new one.

If you decide to rebuild, you will also need carburetor cleaner, a clean work surface (preferably topped with a sheet of white cardboard to enhance seeing removed parts, etc.), and, ideally, a source of compressed air for blowing out loosened debris and solvent. Well-equipped shops might have a small sonic cleaning tank, as also found in some jewelry stores for safely shaking loose tiny bits of dirt from rings, necklaces, and watch mechanisms.

The design of your carburetor depends on the size of the engine and the application. Engines designed for lawn tractors require a precisely tuned carburetor with a choke and idle mixture system. Walk-behind mower engines operate well without these design enhancements. This section offers directions for cleaning and adjusting a range of carburetor types. Yours may look different and may require fewer steps.

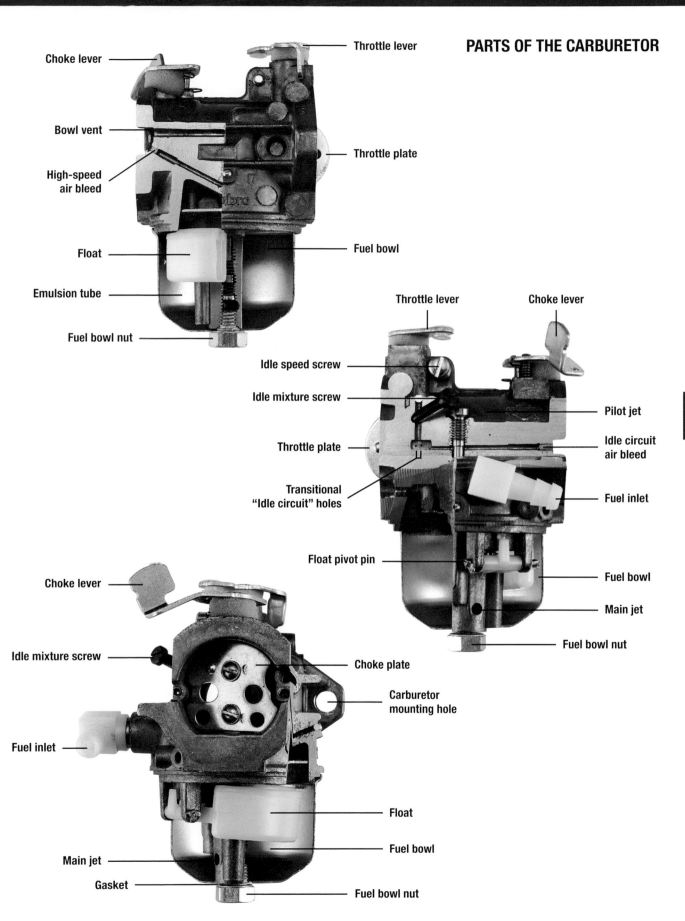

Choke lever

Throttle lever

PARTS OF THE CARBURETOR

Bowl vent

Throttle plate

High-speed air bleed

Float

Fuel bowl

Emulsion tube

Fuel bowl nut

Throttle lever

Choke lever

Idle speed screw

Idle mixture screw

Pilot jet

Throttle plate

Idle circuit air bleed

Fuel inlet

Transitional "Idle circuit" holes

Float pivot pin

Fuel bowl

Main jet

Fuel bowl nut

Choke lever

Idle mixture screw

Choke plate

Carburetor mounting hole

Fuel inlet

Float

Main jet

Fuel bowl

Gasket

Fuel bowl nut

REMOVING THE CARBURETOR

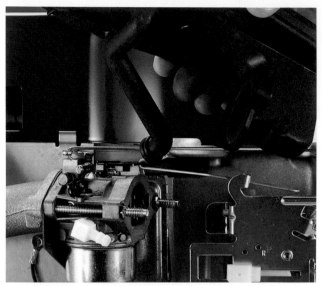

1. Disconnect the spark plug lead and secure it away from the spark plug. Then, remove the air cleaner assembly.

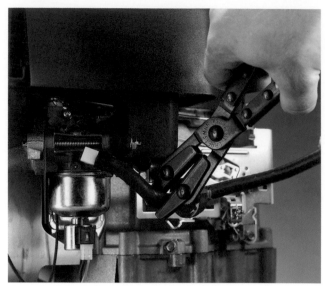

2. Turn off the fuel valve at the base of the fuel tank. If your engine does not contain a fuel valve, use a fuel line clamp to prevent fuel from draining out of the tank while the carburetor is disconnected from the engine.

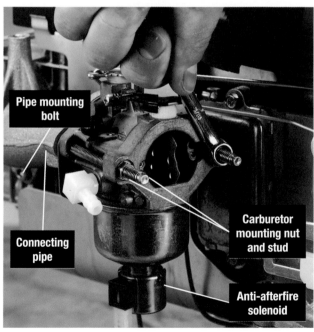

Pipe mounting bolt

Connecting pipe

Carburetor mounting nut and stud

Anti-afterfire solenoid

3. Some carburetors contain an electrical device at the base of the fuel bowl to control afterfire. Disconnect the device, known as an anti-afterfire solenoid, by removing the wire connector from the solenoid's receptacle.

With the carburetor still connected to the governor, unfasten the carburetor mounting bolts. If a connecting pipe joins the carburetor to the engine block, first remove the pipe mounting bolts. Then, disconnect the carburetor from the pipe by removing the nuts and sliding the carburetor off the studs.

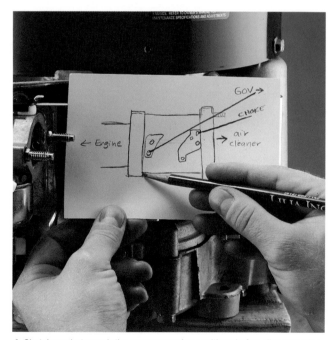

4. Sketch or photograph the governor spring positions before disconnecting them to simplify reattachment. Then, disconnect the governor springs and remove the carburetor, taking special care not to bend or stretch links, springs, or control levers. Linkage removal (from governor or carburetor) can be especially frustrating if not done with a calm attitude and the understanding that it might take a few minutes to decipher the "puzzle" of how the stuff was originally installed. Sometimes the bends needed for successful linkage appear to be one of those, "You can't get there from here," type of conundrums. Even so, a bit of patient study and experimenting almost always produces the solution.

DISASSEMBLING A FLOAT-TYPE CARBURETOR

Your carburetor contains a small amount of fuel. Prepare a clean bowl to catch dripping fuel and store small parts. During disassembly, inspect the bowl for dirt and debris to determine the condition of your carburetor.

ADVANCED REPAIRS

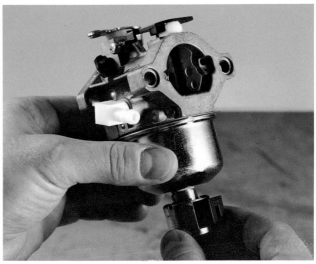

1. Remove the fuel bowl from the carburetor body. The fuel bowl may be attached with either a bolt or the high-speed mixture screw.

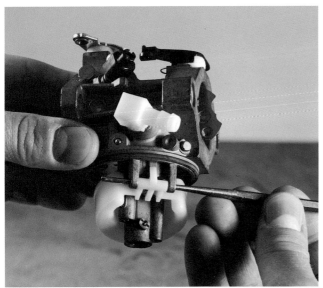

2. Push the hinge pin out of the carburetor body with a small pin or pin punch. Take care to tap only the pin to avoid damaging the carburetor body.

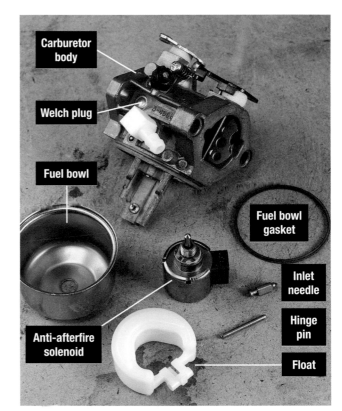

Carburetor body
Welch plug
Fuel bowl
Anti-afterfire solenoid
Fuel bowl gasket
Inlet needle
Hinge pin
Float

3. Remove the float assembly, inlet needle valve, and fuel bowl gasket.

continued

WORKING WITH SOLVENT

Carburetor cleaner is a powerful solvent that can eat up or distort carburetor parts—especially plastic parts—if they are soaked for too long. A carburetor should be soaked for no more than fifteen minutes. Rubber parts, such as seals, O-rings, and pump diaphragms, should never be exposed to carburetor cleaner and should always be removed before soaking.

Use compressed air and carburetor cleaner to clear clogged passages or tiny meter holes in the carburetor. It may be tempting to ream or drill these holes and passages. Never do so. They are precisely sized and may be permanently damaged if any solid material, such as a wire or drill bit, is inserted.

DISASSEMBLING A FLOAT-TYPE CARBURETOR (continued)

4. If your carburetor contains an idle mixture screw, remove it along with the spring.

5. Rotate the throttle plate to the closed position, remove the throttle plate screws and the throttle plate.

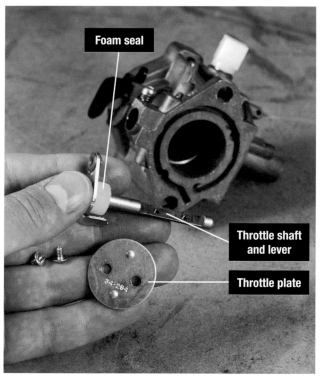

Foam seal

Throttle shaft and lever

Throttle plate

6. Remove the throttle plate shaft and foam seal. Then, remove the choke plate and choke shaft and felt or foam washer in the same manner.

Welch plug

7. Use your carburetor repair kit to identify replaceable welch plugs. These seals cover openings in the carburetor left over from machining. Insert a sharpened $5/32$-inch pin punch at the edge of each plug to be removed and tap cleanly to free the plug.

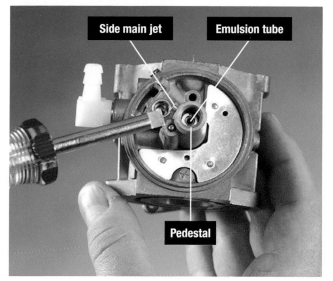

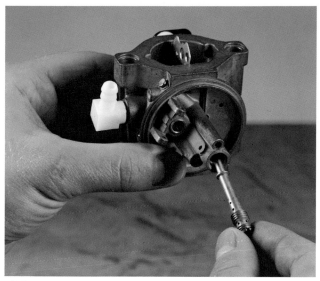

8. Unscrew the main jet from the side of the carburetor pedestal (if equipped). Then, unscrew the emulsion tube; it may be screwed in tight. A carburetor screwdriver is the best tool for the job. It's designed to fit the slot in the head of the emulsion tube so that you won't damage the threads inside the pedestal or the tube itself as you loosen it.

9. Remove the emulsion tube.

INSPECT THE CARBURETOR

Soak metal and plastic carburetor parts in all-purpose parts cleaner for no more than fifteen minutes to remove grit. Or, wearing safety glasses, spray the parts with carburetor cleaner. Then, wipe away solvent and other residue thoroughly, using a clean cloth. Never use wire or tools. They can damage or further obstruct plugged openings.

Inspect all components and use additional carburetor cleaner to loosen stubborn grit and to clear obstructions.

Replace any parts that are damaged or permanently clogged.

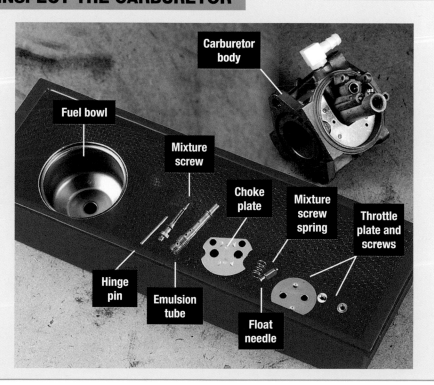

INSPECTING MIXTURE SCREWS

Brass mixture screws control the air-fuel mixture at high speed and at idle. Overtightening can damage the tip of the screw so that proper adjustment is no longer possible (photo A). Remove any nonmetal parts and soak mixture screws in carburetor cleaner for fifteen minutes. Then, inspect them carefully for wear. Replace a mixture screw if the tip is bent or contains a ridge.

COMPENSATING FOR HIGH ALTITUDE

Small engines are set at the factory to operate at average atmospheric pressure. If you live at a high altitude, you may need to modify the carburetor to ensure adequate air or fuel intake. Depending on your model, you will need to remove the small metal fitting near the choke plate, known as the main jet air bleed (photo B) or replace the fixed main jet (photo C) with one designed for high elevations. Ask your authorized service dealer for additional details on the necessary adjustments for your area.

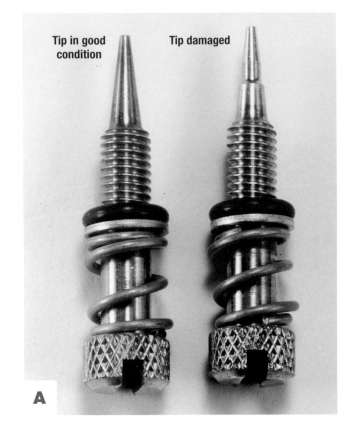

Tip in good condition   Tip damaged

A

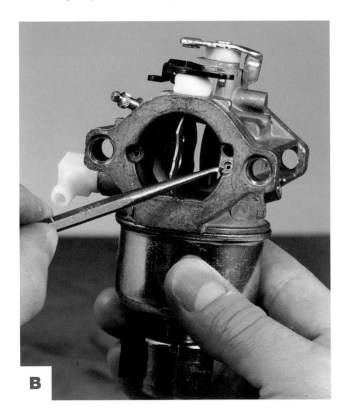

B

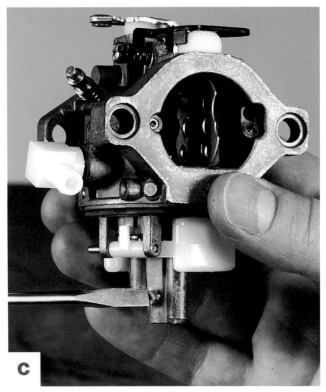

C

REASSEMBLING THE CARBURETOR

Install new welch plugs from your repair kit, using a pin punch slightly smaller than the outside diameter of the plug (photo D). Tap on the punch with a hammer until the plug is flat (strong blows with the hammer will cause the plug to cave in). Then, seal the outside edge of the plug with enamel nail polish.

Assemble the choke by inserting the return spring inside the foam seal and sliding the spring and seal assembly onto the choke shaft. Plastic choke plates have a stop catch at one end of the spring; metal plates have a notch to hold the hook at one end of the spring.

Insert the choke shaft into the carburetor body and engage the return spring. If the choke lever uses a detent spring to control the choke plate position, guide the spring into the notched slot on the choke lever. Place the choke plate on the shaft with the single notch on the edge toward the fuel inlet. Lift the choke shaft and lever up slightly and turn counterclockwise until the stop on the lever clears the spring anchor. Push the shaft down.

D

Insert the choke plate into the choke shaft or attach it with screws so that the dimples face the fuel inlet side of the carburetor. The dimples help hold and align the choke shaft and plate.

Install the throttle shaft seal with the sealing lip down in the carburetor body until the top of the seal is flush with the top of the carburetor. Turn the shaft until the flat side is facing out. Attach the throttle plate to the shaft with the screws so that the numbers on the throttle plate face the idle mixture screw and the dimples face in.

Install the inlet needle seat with the groove down, using a bushing driver. Then, install the inlet needle on the float and install the assembly in the carburetor body.

Insert the hinge pin and center pin. Then, install the rubber gasket on the carburetor and attach the fuel bowl, fiber washer, and bowl nut.

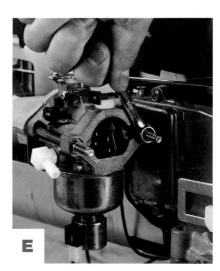

E

ATTACHING THE CARBURETOR AND AIR CLEANER ASSEMBLY

Position the carburetor so the beveled edge fits into the fuel intake pipe and attach the carburetor with nuts or bolts, as required (photo E), leaving these fasteners loose for final tightening with a torque wrench. Consult your authorized service dealer for the proper tightening torque.

Install the air cleaner assembly, making certain that the tabs on the bottom of the air cleaner are engaged.

Inspecting the Flywheel & Key

🔧 **Tools & Materials**

- File
- Flywheel
- Clutch tool
- Flywheel puller
- Flywheel holder
- Socket wrench set

Time required: 1 hour

The flywheel on your small engine was originally designed to store the momentum from combustion to keep the crankshaft turning in between the engine's power strokes (see "Compression System," pages 12 to 13). But the flywheel on today's small engines serves several other purposes. The fins help cool the engine by distributing air around the engine block (see "Lubrication & Cooling System," pages 22 to 24). The fins also blow air across the air vane on a pneumatic governor, maintaining the desired engine speed (see "Governor System," page 25). Magnets mounted in the outside surface of the flywheel are required for ignition (see "Ignition System," pages 20 to 21). And on engines with starter motors, lights or other devices, magnets mounted inside and outside the flywheel are at the heart of the electrical system (see "Electrical System," page 27).

If a lawn mower or tiller blade hits a rock or curb, the flywheel key (left) can sometimes absorb the damage, reducing repair costs significantly. Always check for damage by removing the flywheel to inspect the key and the keyway (the slot on the crankshaft that the key slides into). The soft metal key must eliminate play between the flywheel and crankshaft.

The flywheel key is a small, but important, piece of soft metal mounted between the flywheel and crankshaft to time the engine.

REMOVING THE FLYWHEEL

Disconnect the spark plug lead and secure it away from the spark plug. Then, loosen the bolts holding the shroud in place and remove the shroud.

If the engine is equipped with a flywheel brake, remove any cover and disconnect the outer end of the brake spring (photo A).

If the flywheel is equipped with a flywheel clutch, remove it with a flywheel clutch tool while holding the flywheel with a flywheel holder or a flywheel strap wrench (see "Specialty Tools," page 30). If the flywheel is attached with a nut, use the flywheel holder as a brace, and remove the flywheel retaining nut with the appropriate socket (photo B).

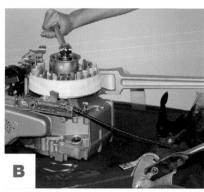

With the flywheel nut threaded onto the crankshaft, install a flywheel puller so its bolts engage the holes adjacent to the flywheel's hub (photo C). If the holes are not threaded, use a self-tapping flywheel puller or tap the holes using a 1/4 × 20 tap. CAUTION: Never strike the flywheel. Even a slightly damaged flywheel presents a safety hazard and must be replaced.

Rotate the puller nuts evenly until the flywheel pops free. Then, remove the flywheel and key.

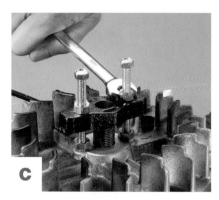

INSPECTING THE FLYWHEEL AND KEY

Check for cracks on your crankshaft or broken fins on the flywheel. Replace them if you find such damage. The tapered sections must be clean and smooth, with no play between the two.

Inspect the keyway and flywheel (photo D) for damage. Slight burrs may be removed with a file. Then, make certain there is no play or wobbling when the flywheel is placed on the crankshaft.

Inspect the flywheel key. If there are any signs of shearing or if you have doubts about the condition of your flywheel key, replace it. It's simple and inexpensive.

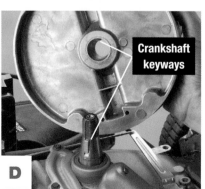

Crankshaft keyways

INSTALLING THE FLYWHEEL

Place the new flywheel that you've obtained from your authorized service dealer on the crankshaft and look through the flywheel hub to align the keyways on the flywheel and crankshaft.

With the flywheel in place, place the key in the keyway; it should fit securely. If you feel play, check to see if the key is upside down. Debris can also prevent the key from seating in the keyway.

Once the key and flywheel are securely in place, reattach the flywheel nut or clutch. Consult your authorized service dealer for the torque specifications for your make and model.

Replacing the Ignition

Tools & Materials

- Bench vise
- Pin punch (3⁄16-inch)
- Razor blade or utility knife
- Shim
- Socket wrench
- Microfiche or index cards (.010 inch)

- Silicone sealer
- Soldering iron
- 60/40 solder
- Spark plug tester
- Wire cutters

Time required: 1 hour

Solid-state ignition armatures have been used for small engines since the early 1980s.

Ignition points were used in small engines prior to the early 1980s. Finding replacement parts has become quite difficult.

Today's small engines contain a solid-state ignition armature mounted adjacent to the flywheel. The only moving parts in the system are the magnets mounted in the flywheel, which interact with the armature to produce electrical current. Most ignition armatures are designed to be replaced, not repaired, if they fail. If yours is an early solid-state ignition armature (say, from the mid-1970s), it may have replaceable parts. But you'll probably find that replacing the armature is the easiest solution if it fails.

Most engines built through the early 1980s contain a set of mechanical points, known as breaker points, under the flywheel. The points open and close an electrical circuit required for ignition. In some cases, you can improve the reliability of such an engine by bypassing the breaker points system using a solid-state ignition retrofit kit. It's an easy modification available at good small-engine shops and auto parts stores.

Before you replace a suspect ignition armature, always test ignition with a spark tester (see "Servicing Spark Plugs," pages 52 to 53). Check for faulty electrical switches that could be the source of the problem (see "Braking System," page 26).

This section covers the procedures for replacing the ignition armature and bypassing a breaker points system with a solid-state system.

INSTALLING AND ADJUSTING A NEW IGNITION ARMATURE

An ignition armature must be set at a precise distance from the flywheel. Ask your authorized service dealer for the proper gap for your engine. Common armature gap ranges are .006-.010 inch and .010-.014 inch. Armatures are often packaged with a shim to assist in setting the gap. Microfiche or index cards of the proper thickness also work well.

Remove the old ignition armature mounting screws (photo A). Disconnect the stop switch wire from the flywheel brake (see "Removing a Brake Pad," page 133) and remove the armature.

Attach a replacement armature from the original engine manufacturer, using mounting screws (photo A). Then, push the armature away from the flywheel and tighten one screw (photo B).

Turn the flywheel so the magnets are on the opposite side from the ignition armature (photo C).

Place the appropriate shim between the rim of the flywheel and the ignition armature. Holding the shim and turn the flywheel until

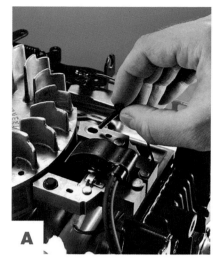

A

B

C

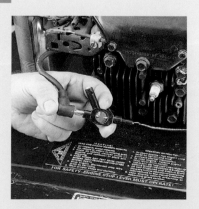

D

the magnets are adjacent to the armature (photo D).

Loosen the tight screw so the magnets pull the ignition

armature against the flywheel and shim. Then, tighten both mounting screws and rotate the flywheel until the shim slips free.

TESTING A STOP SWITCH

Insert the spark plug lead on one end of a spark tester and attach the tester's alligator clip to ground, such as an engine bolt.

Place the equipment stop switch control in the OFF or STOP position. If the engine is not connected to the equipment, ground the stop switch wire to the cylinder. Attempt to start the engine using the rewind cord or key (if equipped). There should be no spark. If a spark appears, inspect the stop switch for damage. Consult your authorized service dealer if you find a faulty switch.

Place the stop switch control in the RUN or START position. If the engine is not connected to the equipment, make sure the stop switch wire is not grounded. Attempt to start the engine. A spark should be visible in the tester. If no spark appears, check for broken wires, shorts, grounds or a defective stop switch.

Confirm that the stop switch is working and reconnect the spark plug lead.

RETROFITTING AN OLDER IGNITION ARMATURE

Breaker point ignition systems were common through the early 1980s. You can improve ignition reliability on a single-cylinder engine equipped with breaker points and a two-leg armature by installing a solid-state ignition conversion kit that bypasses the points. Consult a reputable service dealer for the proper conversion kit.

1. Disconnect the spark plug lead and secure it away from the plug. Then, remove the flywheel and discard the old flywheel key.

A

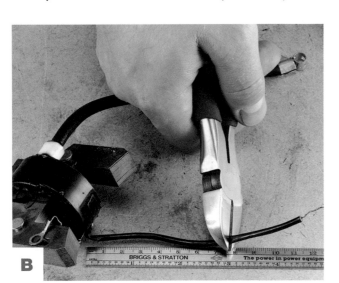

B

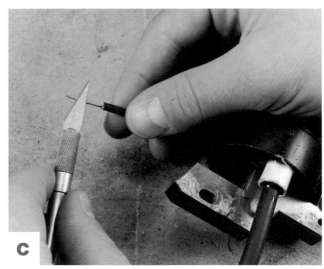

C

2. Cut the armature primary and stop switch wires as close as possible to the dust cover (photo A). Then, remove the dust cover, points, and plunger, and plug the plunger hole with the plug supplied in the conversion kit.

3. Loosen the screws and remove the armature. Then, cut the armature's primary wire to a 3-inch length (photo B). Strip away ⅝-inch of the outer insulation. Then, use a utility knife or razor blade to scrape off thoroughly the red varnish insulation underneath. Take care not to nick or cut the wire (photo C).

D

E

F

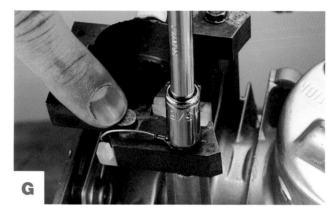

G

4. Install the conversion module (photo D). Modify the air vane brackets or guides for clearance, as required.

5. Fasten a pin punch in a bench vise. Push open the spring-loaded wire retainer by pressing down on the punch. With the slot open, insert the armature's primary wire and a new stop switch wire (if required), together with the module primary wire (photo E). Then, release the wire retainer, locking the wires in place. Secure the wires by soldering the ends with 60/40 rosin core solder.

6. Twist the armature ground wire and module ground wire together (two turns) close to the armature coil (photo F) and solder the twisted section, taking care not to damage the armature coil casing. Avoid crossing these wires with those inserted in the wire retainer in Step 5.

7. Remove the shortest ground wire by cutting it off close to the soldered connection.

Cement the wires to the armature coil, using a generous amount of silicon sealer to protect against vibrations.

8. Use a screw to attach the armature/module ground wire to the armature (photo G). Then, fasten the armature to the engine so that the wire retainer is toward the cylinder.

9. Remove the remainder of the original stop switch wire as close as possible to the terminal on the engine. Then, route the new wire from the module, following the same path as the original. Fasten the new wire in place. Make sure the wire does not interfere with the flywheel.

10. Install the flywheel, using the replacement flywheel key in your kit, and tighten the flywheel nut or rewind clutch (see "Inspecting the Flywheel & Key," pages 108 to 109). Set the armature air gap (see "Installing and Adjusting a New Ignition Armature," page 111). Then, test the stop switches (see "Testing a Stop Switch," page 111).

Testing the Electrical System

An autoranging multimeter is a useful diagnostic tool for troubleshooting the electrical system in power equipment.

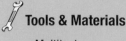

Tools & Materials

- Multitester
- Shim
- Index card or microfiche
- Socket wrench set
- Torque wrench

Time required: 1 hour

Small engines that start with a key require an electrical system to charge the battery and to power on-board electrical devices. If you hear a groan or just a click when you try to start a small engine equipped with an electric starter motor, your electrical system may be the source of the problem. Electrical problems can also keep on-board electrical devices from operating. The proper test can help you identify the source of the problem.

When testing an alternator or other electrical system component, it's imperative that your multitester be connected to the appropriate wires in your engine in order to get pertinent test readings. Because each model of electric starting engine's electrical system may be unique, it is wise to consult the owner's/repair manual or service shop/dealer for details about the proper way to attach a multitester to your engine/alternator.

This section also explains how to replace your stator, if necessary. On most models, the stator is mounted under the flywheel and is not difficult to replace once the flywheel is removed. On some walk-behind lawn mowers, the stator is mounted outside the flywheel, making replacement even simpler.

SELECTING THE RIGHT TEST FOR YOUR ALTERNATOR

With the engine off, locate the thin wire(s) extending from beneath the blower housing. These wires attach to the stator under the flywheel and deliver the electrical current from the stator to the battery and other electrical devices.

Note the color of the wires (scrape away any engine paint to identify the true wire color), as well as the color of the wire connector, typically an inch or two from the blower housing.

Use the appropriate test procedure on page 116.

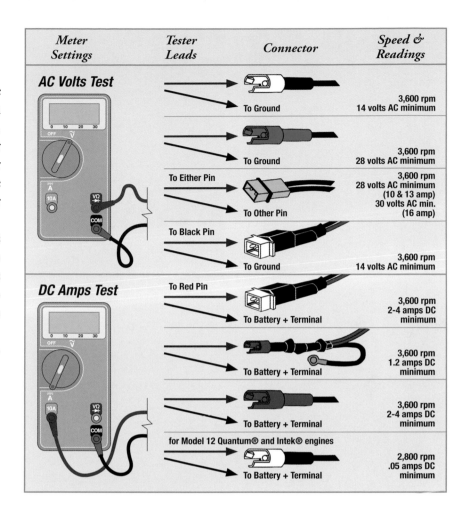

Meter Settings	Tester Leads	Connector	Speed & Readings
AC Volts Test	To Ground		3,600 rpm 14 volts AC minimum
	To Ground		3,600 rpm 28 volts AC minimum
	To Either Pin / To Other Pin		3,600 rpm 28 volts AC minimum (10 & 13 amp) 30 volts AC min. (16 amp)
	To Black Pin / To Ground		3,600 rpm 14 volts AC minimum
DC Amps Test	To Red Pin / To Battery + Terminal		3,600 rpm 2-4 amps DC minimum
	To Battery + Terminal		3,600 rpm 1.2 amps DC minimum
	To Battery + Terminal		3,600 rpm 2-4 amps DC minimum
	for Model 12 Quantum® and Intek® engines / To Battery + Terminal		2,800 rpm .05 amps DC minimum

BATTERY SAFETY

Small engines typically use lead-acid batteries, which store electrical energy using lead plates and sulfuric acid. The electrolyte fluid in the battery loses its sulfuric acid and gains water as the battery is discharged.

Battery electrolyte is extremely corrosive and can cause severe burns to eyes and skin. If spilled onto clothing, it will burn holes in the material. Batteries produce hydrogen gas that can cause an explosion if ignited by a spark or open flame. Minimize safety hazards by observing these precautions.

- Follow the manufacturer's recommended procedure for charging, installation, removal and disposal.
- ALWAYS hold a battery upright to avoid spilling electrolyte.
- Wear protective eyewear and clothing when handling batteries.
- If electrolyte spills on skin or splashes in an eye, flush immediately with lots of cold water and contact a physician immediately.

- Service batteries in a well-ventilated area, away from sources of sparks or flames.

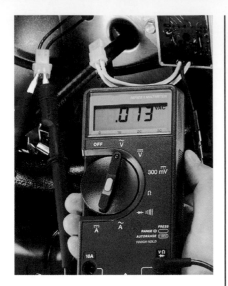

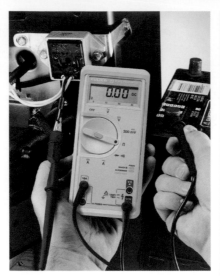

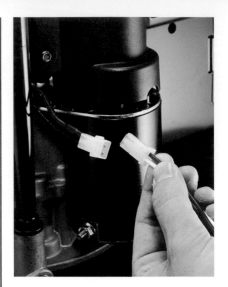

CONDUCTING AN AC VOLTS TEST

If your engine requires an AC VOLTS test, set the tester's dial to AC VOLTS and follow these steps:

Insert the black multitester lead into the tester's COM receptacle. Connect the other end to ground, such as an engine bolt or cylinder fin, or to the double connector on the stator output wires.

Insert the red lead on the multitester into the tester's AC VOLTS receptacle. Connect the red lead to the appropriate stator output wire.

Start the engine and let it run for several minutes to reach its operating temperature. Then, using a tachometer, set the engine test speed and check the reading on the tester. Replace the stator if the reading is incorrect.

Turn off the engine and disconnect the multitester from your equipment.

CONDUCTING A DC AMPS TEST

If your engine requires a DC AMPS test, set the tester's dial to DC AMPS and follow these steps:

Insert the black multitester lead into the tester's COM receptacle. Connect the other end to the battery's positive terminal. (NOTE: The battery must be grounded to the equipment frame or the engine block to create a complete circuit.)

Insert the red lead on the multitester into the tester's AMPS receptacle. Connect the red lead to the appropriate stator output wire.

Start the engine and let it run for several minutes to reach its operating temperature. Then, using a tachometer, set the engine test speed and check the reading on the multitester. An incorrect reading indicates that the stator, diode or regulator should be replaced.

Turn off the engine and disconnect the tester from your equipment.

REPLACING A STATOR UNDER THE FLYWHEEL

In most cases, you need to remove the blower housing, rotating screen, rewind clutch, and flywheel to get to the stator (see "Inspecting the Flywheel & Key," pages 108 to 109). If your stator is mounted outside the flywheel, follow the instructions under "Replacing an External Stator" (opposite).

With the flywheel removed, note the path of the stator wires, under one coil spool and between the starter and starter drive housing.

Remove the ground wire or rectifier assembly (if equipped) from the starter drive housing. Then, remove the stator mounting screws and bushings.

Before installing a new stator, locate the stator wires against the cylinder and make sure the wires remain clear of the flywheel.

Install a new stator assembly, making certain the output wires are properly positioned. While tightening the mounting screws, push the stator toward the crankshaft to take up clearance in the bushing. Then, tighten the screws to 20 inch-pounds.

Reinstall the flywheel, screen and blower housing. Then, attach the ground wire or rectifier assembly (if equipped) to the drive housing.

REPLACING AN EXTERNAL STATOR

Disconnect the stator output wire from wires leading to the battery or other electrical devices.

Rotate the flywheel until the magnets are positioned away from the stator. Then, loosen the stator mounting bolts and remove the stator from the engine.

With the flywheel magnets positioned away from the stator, install the new stator, leaving a wide gap between the stator and flywheel. Tighten one of the mounting bolts.

Reattach the stator output wires. Then, follow the procedure for "Adjusting the Air Gap on an External Stator" (below).

ADJUSTING THE AIR GAP ON AN EXTERNAL STATOR

The gap between the stator and the flywheel must be set precisely for the stator to function properly. Many stators require a .010-inch stator air gap. Consult your authorized service dealer for the proper gap for your stator.

Rotate the flywheel until the magnets are positioned away from the stator.

Loosen both stator mounting bolts and move the stator away from the flywheel. Then, tighten one of the mounting bolts.

Place a shim or microfiche card of the proper thickness between the stator and the flywheel (photo A).

Turn the flywheel until the magnets are adjacent to the stator.

Loosen the tightened bolt and let the magnets pull the stator until it is flush with the shim.

Tighten both mounting bolts to 25 inch-pounds (photo B).

Turn the flywheel while pulling on the shim to release it.

A

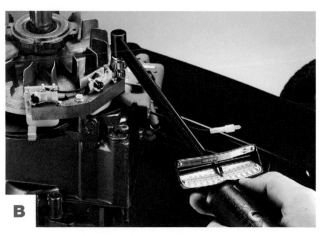

B

Tools & Materials

- Cylinder head replacement gasket
- Nylon-faced hammer
- Putty knife
- Rubber gloves
- All-purpose solvent
- Steel wool
- Torque wrench
- Wire brush
- Wooden or plastic scraper

Time required: 2 hours

Carbon deposits on critical engine parts, such as the cylinder head above, have a negative impact on engine performance and should be cleaned off every 100 hours or so.

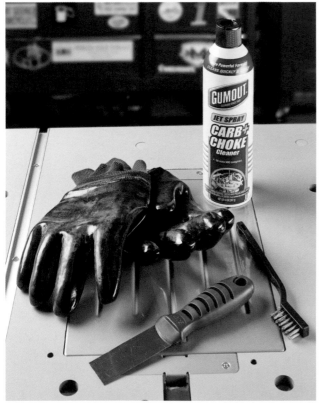

All rubber components and some plastic parts are subject to distortion if carb cleaner is sprayed directly on them.

Removing Carbon Deposits

One byproduct of combustion is carbon, the black soot that can collect and harden on the cylinder head, cylinder wall, piston, and valves. Carbon deposits in the combustion chamber can affect engine performance, resulting in higher oil consumption, engine knocking, or overheating.

It is prudent to remove the cylinder after each 100 hours of operation and scrape off the carbon, using the tools and solvents described in this section. Clean the cylinder more frequently if you use your engine heavily.

REMOVING ENGINE COMPONENTS

The first step in servicing the cylinder head is reaching the cylinder head. You may need to remove some other components first.

Remove the muffler, muffler guard, and any other components that block access to the cylinder.

Cylinder head bolts near the muffler and exhaust port may be longer. To avoid subsequent confusion of which bolt goes where, prepare a "keeper" template. Draw a rough outline of the cylinder head on a piece of cardboard and punch holes for each bolt location. Then, remove the cylinder head bolts and insert them through the corresponding holes in this "keeper" template until you are ready to reinstall the cylinder head (photo A). In an old or neglected engine, cylinder head bolts may be rusted or corroded and difficult to remove without breakage. A broken one requires drilling and backing out with a stud extractor tool. Often, this also necessitates cleaning up the threads in the cylinder block.

Lift off the cylinder head. If the head sticks, strike it on the side with a nylon-faced hammer. This should loosen the cylinder head enough for you to gently lift it off the engine. NOTE: Do not pry off the cylinder head. This can damage the surface of the engine block or the cylinder head.

Remove and discard the old head gasket (photo B).

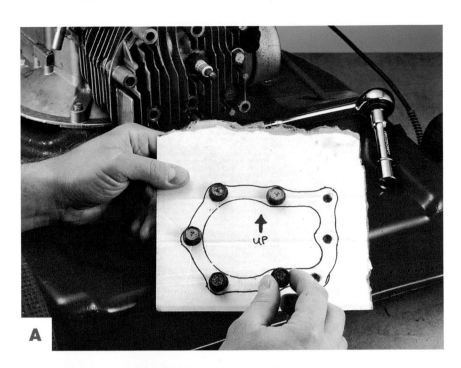
A

B

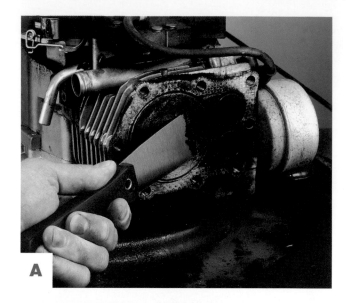

A

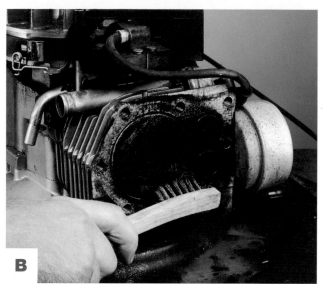

B

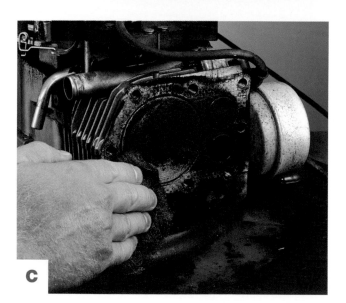

C

REMOVING CARBON DEPOSITS

Always wear protective eyewear and solvent-proof gloves when removing carbon. Ask your authorized service dealer to recommend an all-purpose solvent that will not harm aluminum or plastic components or leave unwanted residues.

Place the piston at top dead center so that the valves are closed. Then, scrape carbon gently from the cylinder head, using a wooden or plastic scraper. Take care not to dig the scraper into the aluminum. On stubborn deposits, use a putty knife, wire brush, or steel wool (photos A, B, and C), taking care not to bear down on the metal surfaces.

Clean away the remaining carbon with solvent, using fine steel wool to smooth rough spots. You can also soak metal parts for up to fifteen minutes to remove stubborn deposits. Scrape again, if necessary, to loosen stubborn grit. Then, clean the area thoroughly with the solvent and set the head aside.

With the piston still at the top of the cylinder and the valves closed, use the same method to remove carbon deposits from the piston and the end of the cylinder (photo C). Note that pistons in most 2-stroke engines have an exhaust deflector "crown" and are not as easy to scrape clean as a flat 4-stroke piston. Some repair shops' policy is to remove shaped pistons for cleaning on a wire wheel.

D

E

Turn the crankshaft to open each valve, and carefully remove any visible carbon deposits on the valves and valve seats (photo D), using only a brass wire brush. CAUTION: Do not allow grit to fall into the valve chambers or between the piston and the cylinder wall (photo E).

Inspect the valves and valve seats to see if they are cracked, rough, or warped. Bring damaged parts to an authorized service dealer for inspection before reassembling the head.

Using a scraper, solvent, or both, remove any remaining carbon and residue left behind by the head gasket on the cylinder head and engine block. Clean the surfaces thoroughly before installing the new head gasket. Any debris or oil left on the cylinder head or engine block may prevent a tight seal and cause eventual engine damage.

REASSEMBLING THE CYLINDER HEAD

Inspect the surfaces of the engine block, cylinder head, and new head gasket to be sure they are clean.

Place the new head gasket in position on the engine block. Do not use sealing compounds.

Set the cylinder head on the head gasket, aligning the cylinder head with the gasket and the engine block.

Remove each head bolt from its hole in the cardboard "keeper" template. Removing any rust, corrosion, or carbonized dirt from them with a wire brush at this time is a good practice. Then, prior to reinstalling the bolts through the cylinder head and into the block, lightly oil them and reinsert the bolts into position using finger-turning, at first, to avoid cross-threading. Leave each bolt a bit loose and ready for final tightening with a wrench. Make sure to attach any housings or brackets that are held in place by the head bolts.

Using a wrench, tighten the cylinder head bolts until each one is just snug. This is best accomplished in a "criss-cross" fashion where, after tightening a bolt on one side of the head, you next tackle the one opposite it in order to keep an even pressure on the mating surfaces. For final tightening, use a torque wrench. Proceed in

increments of roughly one-third the final torque. Consult your owner's manual for the final torque "criss-cross" sequence and torquing specifications. Avoid tightening a single bolt all the way before tightening the other bolts, otherwise this uneven tightening is likely to warp the cylinder head.

Servicing the Valves

Tools & Materials

- Feeler gauge set
- Hex wrench set
- Needlenose pliers
- Nut driver set
- Calipers
- Safety eyewear

- Socket wrench set
- Torque wrench
- Lapping compound
- Valve lapping tool
- Valve spring compressor

Time required: 90 minutes

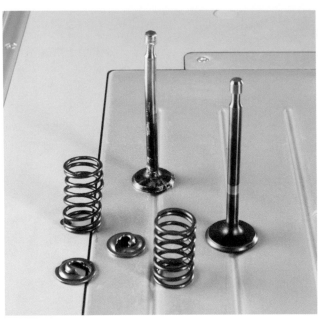

Small-engine valves are not big but their job is very important. A typical assembly includes the valve, valve spring, valve rotator, and valve spring retainer.

A valve spring compressor is an essential tool for removing and installing valves on L-head engines and some overhead valve engines, both of the 4-stroke type.

Valves control the flow of fuel vapor into the combustion chamber and the flow of exhaust gases leaving the engine. Faulty or dirty valves may stick and can develop pits, cracks, or grooves that cause the engine to lose power and fuel efficiency.

When you remove the valves from the engine, inspect them carefully. Then, if the valves are not badly worn and the parts are not damaged, you can tune up the valves and seats (see "Lapping the Valves," page 127) so that the valves seal effectively.

Valves contain a stem, neck, head, and face. Each valve stem moves in a valve guide that is machined directly in the cylinder block or in a replaceable bushing. Each valve also moves through a valve spring, adjacent to the guide, that pushes the valve toward the closed position and holds the valve face against the valve seat. Each valve spring is held in place by a valve spring retainer. Some valve assemblies also include a rotator, a circular component that turns the valve slightly in each cycle to ensure a symmetrical wear pattern on valves and seats. Valves are opened by tappets that ride on the camshaft inside the crankcase.

This section covers the procedures for removing, inspecting, cleaning, and replacing valves and related parts.

VALVE DESIGN

Valve design for the 4-stroke small engine includes one intake valve and one exhaust valve per cylinder. The diagrams on this page show the parts of typical valves in detail and their locations in the engine.

Intake valves open to allow the air-fuel mixture to enter the combustion chamber. Exhaust valves open to allow spent fuel gases to leave the engine. Both valves close to seal the combustion chamber for the piston's compression stroke.

Valve springs push the valves toward the closed position, so that they open only at precisely timed intervals. The valves are pushed open by tappets that ride on the lobes of the camshaft. The camshaft turns along with the crankshaft; both are driven by the movement of the piston. This synchronizes the actions of the valves with those of the piston.

In L-head engines, the valves are located to one side of the cylinder. The valve stems run through the cylinder block, parallel to the piston.

In overhead valve engines, the valves are located in a cylinder head that is much larger than that found in an L-head engine. Overhead valves are pushed open by pivoting rocker arms, operated by push rods. The push rods, in turn, are pushed toward the rocker arms by the tappets. The slightly more complex design yields greater power.

L-Head

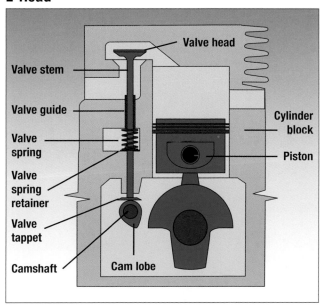

Overhead valve (OHV)

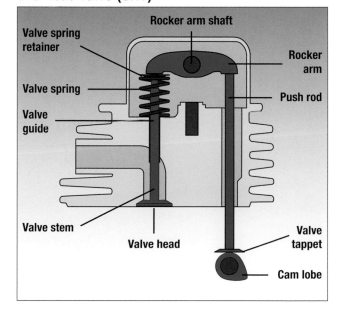

MACHINING VALVES

The valves on your 4-stroke engine may still function after years of service— with no bent, cracked, or damaged parts. But in all likelihood, they are worn and no longer form a good seal when the valves are closed. The result is loss of power and wasted fuel each time you run your engine.

While a few do-it-yourselfers obtain valve grinding stones that can be activated via an electric drill, most small-engine experts agree that it is wiser for all but the experienced amateur mechanic to take his/her engine to a professional shop where a technician can routinely restore the valve seats or install ones. This pro will also have quick access to any required replacement parts, such as valve seat inserts.

A

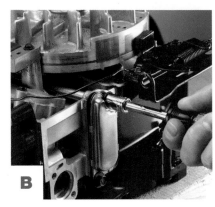

B

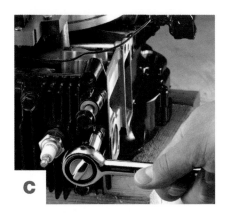

C

REACHING THE VALVE CHAMBER

Before you can service the valves, remove engine components that interfere. NOTE: Always wear safety eyewear when removing and installing valves.

Remove the muffler (see "Inspecting & Changing the Muffler," pages 92 to 95), crankcase breather (see page 59), and any other components that block access to the valve chamber (photos A and B).

Remove the cylinder head bolts (photo C). Label the bolts, if necessary, to ensure proper installation later, since they may be of different lengths (see "Removing Carbon Deposits," pages 118 to 121).

VALVE SPRING COMPRESSOR

If you plan to remove the valves from your 4-stroke engine, a valve spring compressor is a must. Removing valves without this tool is unsafe and can be an exercise in frustration. The method for using the tool varies depending on the type of valve assembly and the design of the engine block. Some valve assemblies hold the valve spring in place with a pin or a pair of collar-shaped automotive-type retainers. Others use a retainer with a keyhole-shaped slot that locks onto the valve stem. No matter which type of retainer you find on your valves, a valve spring compressor allows you to do the job right.

REMOVING THE VALVES
(AUTOMOTIVE TYPE OR PIN RETAINERS)

Adjust the jaws of the valve spring compressor until they touch the top and bottom of the valve chamber.

Push the tool in until the upper jaw slips over the upper end of the spring. Tighten the jaws to compress the spring (photo D).

Remove the retainers and lift out the valves, compressors and springs.

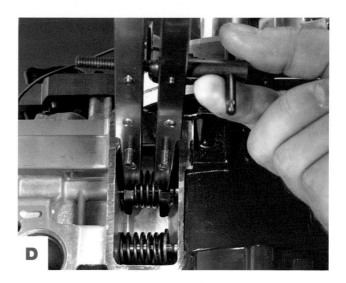

REMOVING THE VALVES
(KEYHOLE RETAINERS)

Removing keyhole retainers requires some patience. Keep in mind the retainer's key-shaped slot. This will help you slip the retainer off the valve stem, even when the retainer is hidden from view by the valve spring compressor.

1. Slip the upper jaw of the valve spring compressor over the top of the valve chamber and the lower jaw between the spring and retainer. If the engine design does not permit the upper jaw to fit over the top of the valve chamber, insert the upper jaw into the chamber over the top of the spring, so that the spring is between the tool's jaws (photo E).

2. Rotate the handle on the valve spring compressor clockwise to compress the spring. Then, slide the retainer off the valve by shifting it with needlenose pliers so that the large part of the keyhole is directly over the stem. Use the pliers to remove the retainer from the valve chamber (photo F).

3. With the valve spring compressor clamping the spring, remove the tool and spring from the chamber. Then, slowly crank open the valve spring compressor to release the tension and remove the spring.

INSPECTING THE VALVES

Before wiping or cleaning the valves, look them over carefully. Residue on the valves may help you identify a specific problem. Gummy deposits on the intake valve go hand in hand with a decrease in engine performance, often because the engine has been run on old gasoline. Hard deposits on either valve suggest burning oil, which has several possible causes (see page 39). Follow the steps below to check for the most likely sources of valve problems.

1. Check the valve face for an irregular seating pattern. The pattern around the face should be even with the valve head and of equal thickness all the way around. Then, look for stubborn deposits. Remove them with a wire brush and solvent, soaking the parts for several hours, if necessary, to loosen hardened grit.

2. Run a fingernail or credit card along the valve stem once you have cleaned it. If you feel a ridge, the valve stem is worn and should be replaced. Keep in mind that the valve guide may also be worn and need replacement by a machinist (see "Machining Valves," page 124).

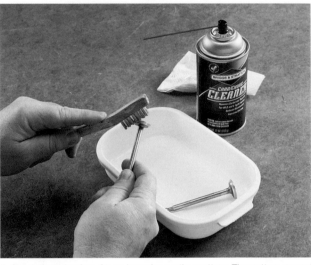

3. Measure the thickness of the valve head, known as the valve head margin, using a caliper. Replace the valve if the margin measures less than 1/64 inch.

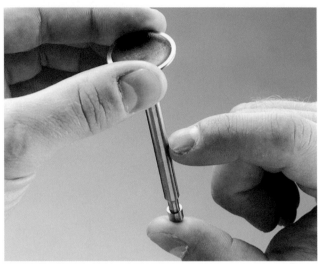

4. Examine the surfaces of the valve face and seat. An uneven wear pattern tells you it's time to replace them both or resurface the seat and replace the valve (see "Machining Valves," page 124). Check that both valve springs are straight. Replace either spring if it is bent.
NOTE: The exhaust valve spring may use thicker wire than the intake valve spring.

LAPPING THE VALVES

You can smooth out minor scoring and pitting of the valve face and seat and restore a valve's ability to seal the combustion chamber by lapping the valves. This procedure involves rotating the valve in the seat with a layer of lapping compound in between the valve and seat. A lapping tool is used to hold and rotate the valve. During lapping, you'll need to check your progress often. Otherwise, it is easy to remove not only the carbon buildup, but also much of the metal, further damaging the valve or seat.

Apply a small amount of valve lapping compound to the valve face and insert the valve into the valve guide. Wet the end of the lapping tool suction cup and place it on the valve head. Spin the valve back and forth between your hands several times. Lift the tool, rotate one-quarter turn, and spin again.

Clean the surface frequently and check your progress. Lap only enough to create a consistent and even pattern around the valve face.

Once lapping is completed, clean the valves thoroughly with solvent to ensure that all of the abrasive residue is removed. Any particles that remain can rapidly damage the valves and other engine components.

ADJUSTING TAPPET CLEARANCES

Since lapping removes a small amount of material from the surfaces of the valve face and valve seat, you may need to adjust the tappet clearances—the spacing between the valve stem and the tappet—after lapping and reinstalling the valves. Consult a shop manual or ask your authorized service dealer for the correct tappet clearance for your engine.

With each valve installed in its proper guides in the cylinder, turn the crankshaft (clockwise as viewed from the flywheel end of the crankshaft) to top dead center. Both valves should be closed. Then, turn the crankshaft past top dead center until the piston is ¼ inch down from the top of the cylinder.

Check the clearance between each valve and its tappet, using a feeler gauge. If clearance is insufficient, remove the valve and grind or file the end of the valve stem square to increase the clearance. Check the length frequently as it is easy to remove too much metal.

Once the individual valve parts have been thoroughly cleaned, lubricate the valve stems and guides, using valve guide lubricant. Then, make certain there is NO lubricant on the ends of the valve stems or tappets.

VALVE LIFE

The life of a standard exhaust valve is often shortened because of burning, which occurs when combustion deposits lodge between the valve seat and the valve face. These deposits prevent valves from closing and sealing completely. Deposits left over from burning are more common on engines that operate at constant speeds and constant loads for long periods, such as generators.

Valve life can be extended using a valve rotator, which turns the exhaust valve slightly on each lift, wiping away deposits lodged between the valve face and seat. Exhaust valves which can be used together with valve rotators have a greater resistance to heat, so can reduce this problem.

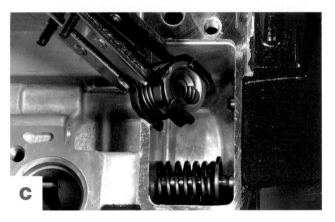

REINSTALLING VALVES WITH KEYHOLE RETAINERS

You need safety eyewear during this step, to protect yourself from the possibility of a flying spring.

Valves with keyhole retainers do not require an additional retainer. Compress the keyhole retainer and spring with the compressor tool—the large hole should face the opening in the tool—until the spring is solid (photo A).

Brush the valve stem with valve stem lubricant (photo B).

Insert the compressed spring and retainer into the valve chamber (photo C).

Insert the valve stem through the large slot in the retainer (photo D). Then, push down and in on the valve compressor until the retainer bottoms out on the valve stem shoulder.

Reinstall the crankcase breather and other components.

INSTALLING VALVES WITH PINS OR AUTOMOTIVE-TYPE RETAINERS

Once again, safety eyewear is absolutely necessary. Remember: a spring that is under tension can and often decides to pop loose and fly through the air.

Place the valve spring into the valve spring compressor and rotate the tool's handle until the spring is fully compressed.

Insert the compressed spring into the valve chamber.

Brush the valve stem with valve stem lubricant (photo B). Then, lower the valve stem through the spring (photo D). Hold the spring toward the top of the chamber and the valve in the closed position.

If pins are used, insert each pin with needle-nose pliers. If automotive-type retainers are used, place the retainers in the valve stem groove.

Lower the spring until the retainer fits around the pin or automotive-type retainer. Then, pull out the valve spring compressor.

Reinstall the crankcase breather and other components.

REMOVING OVERHEAD VALVES

Overhead valve designs vary from one 4-stroke engine model to another. The parts and servicing steps in your overhead valve cylinder may differ from the approach that follows, which is based on a 6-horsepower American-made engine that does not require the use of a valve spring compressor, making valve removal and installation simple.

A

B

C

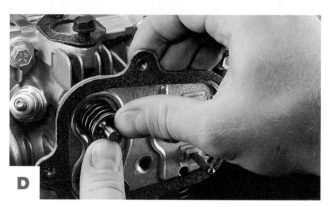

D

ADVANCED REPAIRS

Remove the air cleaner assembly, fuel tank, oil fill tube, blower housing and rewind starter, muffler guard, muffler, carburetor, and any other parts that block access to the cylinder head.

Remove the screws from the valve cover, using a socket wrench or nut driver (photo A). Then, remove the valve cover, breather valve assembly (if equipped), and any gaskets.

Remove the rocker arm bolts with a socket wrench or nut driver (photo B). Then remove the rocker arms and push rods.

Remove the valve caps (if equipped). They are seated on the valve stems (photo C).

Use your thumbs to press in on the spring retainer and valve spring over one of the valves. With the valve spring compressed, remove the retainer

(photo D). If your engine uses a keyhole retainer, line up the large slot in the retainer with the valve stem and release the spring slowly so that the stem slips through the large slot. Then, repeat the procedure for the other valve.

Remove the push rod guide bolts and push rod guide.

Remove the cylinder head bolts and remove the cylinder head by rocking it with your hands. If necessary, loosen the cylinder head by striking it with a nylon-faced hammer. Never pry it loose, as this may damage the head.

Remove and inspect the valves, guides and seats (see "Inspecting the Valves," page 126). The intake and exhaust valves often are made of different steel alloys and may be different colors.

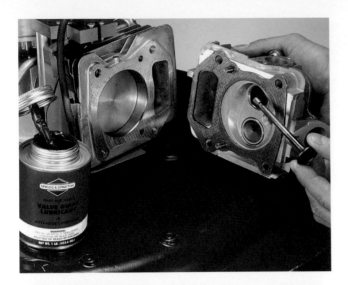

INSTALLING OVERHEAD VALVES

1. Check that valve stems and guides are free of debris and burrs. Then, lightly coat the valve stems with valve guide lubricant and insert them in the cylinder head, taking care to place the correct valve in each valve guide.

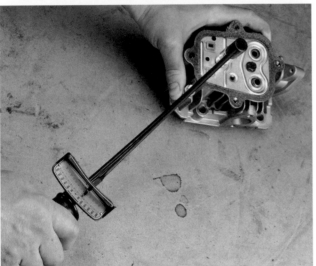

2. Place the push rod guide on the cylinder head and attach the mounting bolts, using a torque wrench. Coat the rocker arm stud threads with a hardening sealant and install the rocker arm studs, using a socket wrench. Consult your authorized service dealer for the proper torque settings for the mounting bolts and studs.

   Lubricate the inside diameter of each valve stem seal (if equipped) with engine oil and install the seals on the valve stems. Press them into place.

3. Install a valve spring and retainer over each stem. Use both thumbs to compress the spring until the valve stem extends through the large end of the keyhole slot. Check that the retainer is fully engaged in the valve stem groove. Repeat this step for the other valve.

   Coat the threads of the cylinder head bolts with valve guide lubricant. Install a new cylinder head gasket on the cylinder, insert the bolts in the cylinder and position the cylinder head on the cylinder.

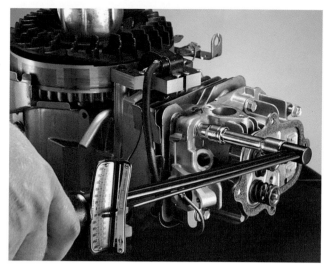

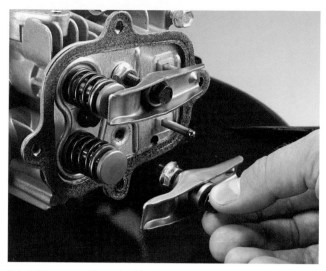

4. Tighten the cylinder head bolts in increments, using a torque wrench. Turn each bolt a few turns, then proceed to the next bolt until each bolt is just snug. Then, for final tightening, proceed in increments of roughly one-third the final torque. Consult your owner's manual for final torque specifications. Uneven tightening is likely to warp the cylinder head.

Install the push rods through the push rod guides and into the tappets.

5. Install the caps on the ends of the valves and wipe away any lubricant. Then, install the rocker arm assemblies while holding the rocker arms against the valve cap and push rod.

Rotate the flywheel at least two revolutions to be sure the push rods operate the rocker arms.

ADJUSTING OVERHEAD VALVES

1. Release the brake spring. Then, turn the flywheel to close both valves.

Insert a narrow screwdriver into the spark plug hole and touch the piston. Turn the flywheel clockwise past top dead center until the piston has moved down ¼ inch. Use the screwdriver to gauge the piston's range of motion.

2. Check the valve clearance by placing a feeler gauge between the valve head and the rocker arm. Clearances differ for the two valves and typically range from .002-.004 inch to .005-.007 inch. Ask your authorized service dealer for the proper valve clearances for your make and model.

Adjust the clearances as required by turning the rocker screw. Once adjustments are completed, tighten the rocker nut.

Install the valve cover, using new gaskets, as required, and make sure the cover is secure.

ADVANCED REPAIRS

Servicing the Brake

Tools & Materials

- Caliper
- Multitester
- Needlenose pliers
- Socket wrench set
- Starter clutch adapter (for some models)
- Tang bending tool
- Torque wrench

Time required: 1 hour

The braking system on a small engine employs a caliper and pad to stop the flywheel when the brake is activated.

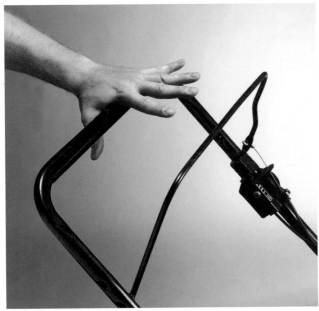

Most small-engine equipment is equipped with a brake bail for your safety. The brake bail is designed to protect you by stopping the engine and any cutting equipment any time you release it. Temptations to tie-wrap, tape, or velcro the brake bail to the mower handle should be strongly resisted!

A well-maintained braking system should stop the engine and any attached cutting equipment within three seconds whenever you step away from the equipment or release the brake bail. A brake bail is standard equipment on today's mowers, tillers, and other walk-behind equipment. The stop switch immediately grounds the ignition, shutting off the engine, while a brake pad or band stops the flywheel from spinning.

If the engine operates for more than three seconds after the bail is released, the stop switch may be faulty. If the blade spins for more than three seconds, the brake pad or band may be worn or in need of adjustment. Many 21st Century-built models use a brake pad that requires no adjustment. Some older models use a brake band, which may require adjustment by an authorized service technician. This section covers the replacement procedures for the brake band and pad styles.

REMOVING A BRAKE PAD

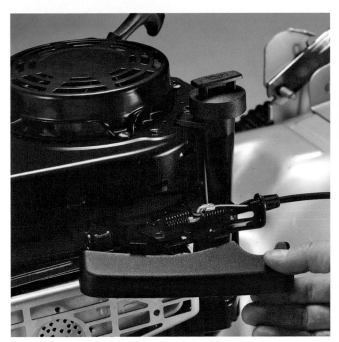

1. Remove the spark plug lead and secure it away from the spark plug. Then, remove any other components that block access to the brake, such as the finger guard, fuel tank, oil fill tube (left), blower housing, or rewind starter (right).

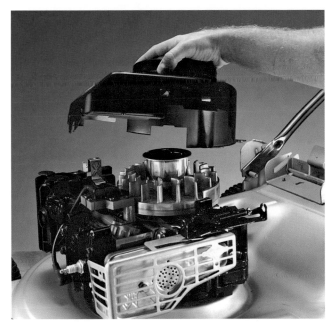

2. Remove the brake control bracket cover, if equipped. Then, loosen the cable clamp screw and remove the brake cable from the control lever.

3. Disconnect the spring from the brake anchor, using needlenose pliers. Then, remove the stop switch wire from the stop switch by gently squeezing the switch and pulling lightly on the wire until it slips free. If the engine is equipped with an electric starter motor, disconnect the pair of wires leading to the starter motor. Loosen the brake bracket screws and remove the bracket from the brake assembly.

INSPECTING AND TESTING A BRAKE PAD SYSTEM

Inspect the brake pad for nicks, cuts, debris, and other damage. Check for wear, by measuring the pad's thickness with a ruler or caliper (photo A). NOTE: Measure the pad only, not the bracket. Replace the brake assembly if the pad's thickness is less than .090 inch.

Test the stop switch, using a multitester or ohmmeter, to determine whether the ignition circuit is grounded when the stop switch is activated (photo B). The stop switch should show continuity (0 ohms) to engine ground when the switch is set to STOP, and no continuity (∞), when the switch is set to RUN. If you identify a problem, check for loose or faulty connections.

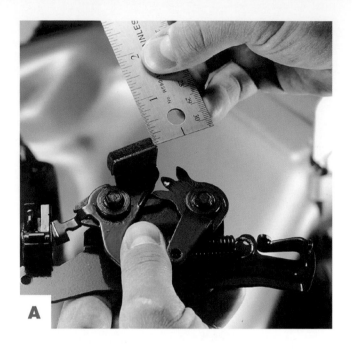

A

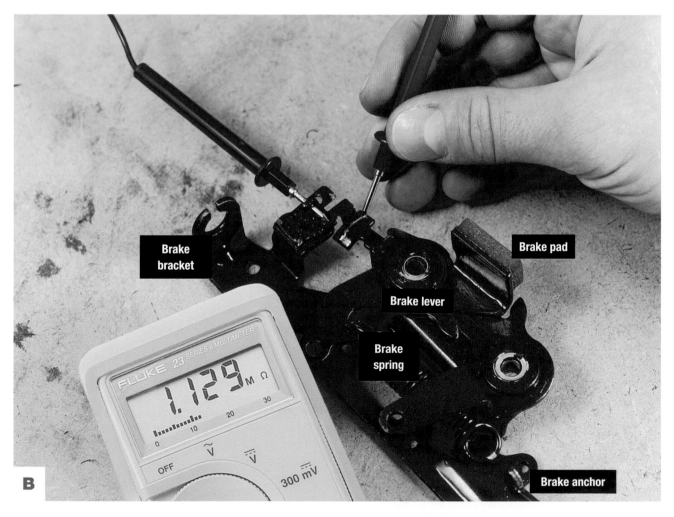

Brake bracket

Brake pad

Brake lever

Brake spring

Brake anchor

B

REASSEMBLING THE BRAKING SYSTEM

Install the brake assembly on the cylinder (photo C). Tighten the mounting bolts to 40 inch-pounds, using a torque wrench.

Install the stop switch wire, bending the end of the wire 90 degrees (photo D).

Install the blower housing and any other engine components removed for brake servicing.

Check the braking action by pivoting the lever. Make sure the lever moves freely and the pad makes full contact with the flywheel.

Attach the brake spring, using needlenose pliers, and connect the brake cable that connects to the brake bail on your walk-behind equipment.

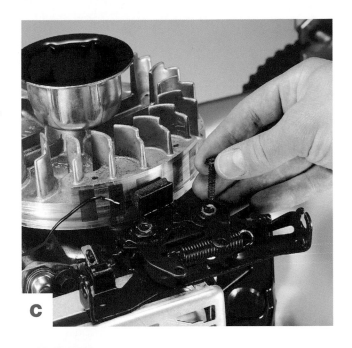

C

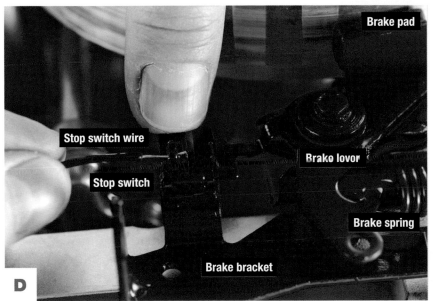

Brake pad

Stop switch wire

Brake lever

Stop switch

Brake spring

Brake bracket

D

Test the braking system by starting the engine and then releasing the brake bail. The engine and the blade or other equipment should come to a stop within three seconds. If you are uncertain about the effectiveness of your braking system, bring the equipment to your authorized service dealer for further inspection.

BRAKE SAFETY

The only safe way to use the brake bail on your small-engine equipment is to pull and hold the bail by hand when starting and running the engine. You can release it when necessary to stop the engine. Keeping the bail in the operating position by any other means overrides an important safety mechanism. The bail is required by law on any mower sold after the early 1980s and is designed to protect you from injury.

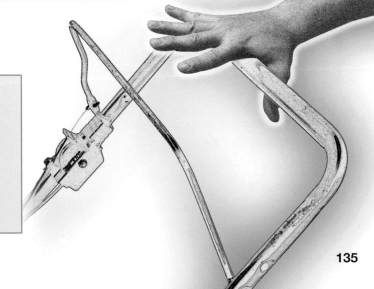

REMOVING AND INSPECTING A BAND BRAKE

The brake band contains loops at either end, mounted on a stationary and a movable post. A tang over the movable post prevents the brake band from dislodging during operation.

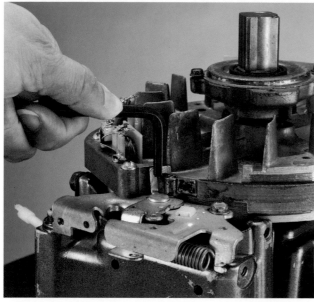

1. Use a tang bending tool (see "Tang Bending and Other Adjustment Methods," pages 88 to 89) to bend the control lever tang outward so it clears the band brake loop.

2. Release the brake spring, using pliers.

3. Lift the band off the stationary and movable posts. Inspect the band for damage. Replace it if you find nicks or cuts.

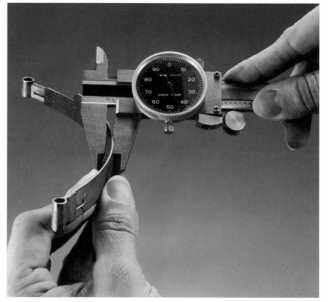

4. Check for wear, by measuring the pad's thickness with a ruler or caliper. NOTE: Measure the pad only, not the metal band. Replace the brake band if the pad's thickness is less than .030 inch.

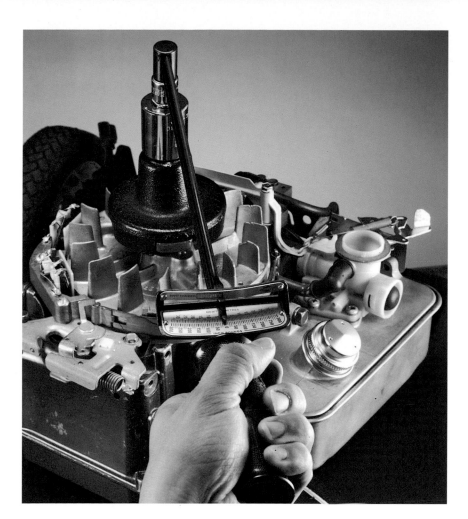

ASSEMBLING A BAND BRAKE

Reinstall the stop switch wire on the control bracket. On older systems, reinstall the stop switch wire on the control bracket stop switch terminal.

Place the band brake on the stationary post and hook it over the end of the movable post until the band bottoms out. NOTE: The brake material on a steel band must be on the flywheel side after assembly. On older systems, install the band brake on the stationary and movable posts.

Bend the retainer tang until it is positioned over the band brake loop so that the loop cannot be accidentally dislodged. After assembly, check that the braking material on the metal band faces the flywheel.

TESTING A BAND BRAKE

Test the band brake's stopping power with the spark plug lead secured away from the spark plug. On electric start engines, disconnect and remove the battery.

With the brake engaged, turn the starter clutch, using a starter clutch adapter and torque wrench. Turning the flywheel clockwise at a steady rate should require at least 45 inch-pounds of torque. If the torque reading is lower, components may be worn, damaged, or in need of adjustment.

Test the stop switch, using a multitester or ohmmeter, to determine whether the ignition circuit is grounded when the stop switch is activated. The stop switch should show continuity (0 ohms) to engine ground when the switch is set to STOP, and no continuity (∞) when the switch is set to RUN. If you discover a problem, check for loose or faulty connections.

2-Stroke Power Equipment

In a 2-stroke engine, the piston acts as a valve, exposing the intake and exhaust ports at designated moments in the cycle. Two-stroke engines are still widely used for chain saws, leaf blowers, and other hand-held equipment. They are also common for outboard motors and motocross motorcycles. They are no longer used on street bikes in many countries because of older 2-stroke engines' higher emissions.

Two-stroke engines have long been preferred for hand-held equipment because of their lightweight design. The same five events (intake, compression, combustion, power, and exhaust) that occur in a 4-stroke engine occur using fewer parts. However, the latest technology has reduced the weight of 4-stroke engine components, creating the potential for inroads in the hand-held equipment industry.

Technological advances in gas and oil injection/computerized ignition have also benefited 2-stroke engine design, with some contemporary outboard motors, for example, burning their fuel and recombusting related exhaust so completely that resulting emissions are less than that of their 4-stroke counterparts.

2-STROKE THEORY

Two-stroke power plants are simpler, have fewer moving parts, and are often lighter than their 4-cycle sisters. They generate power on every two strokes (one up to compress and ignite the fresh fuel/air and then, pushed by the resulting explosion, one down to uncover ports for exhaust escape and other ports to let in a fresh charge of fuel/air) of piston travel.

Many 2-stroke engines have a reed valve between the carburetor and crankcase that acts as a tiny safety door—operated by crankcase vacuum pressure—letting fresh fuel/air into the crankcase to await its subsequent entry through the cylinder ports, while preventing any fuel/air from escaping. In a 2-stroke motor, crankcase pressure is especially important. That's why all of the engine's requisite seals and mating surfaces must be as airtight as possible. While a 4-stroke with bad seals and poorly mated crankcase surfaces will leak oil and rattle (though run in some fashion), its 2-stroke sibling in such a condition probably won't even do more than utter a few pops while the starter rope is yanked.

When healthy, the average 2-stroke will start quicker (because it requires only one revolution of the crankshaft to generate power) and yield higher crankshaft revolutions-per-minute, though lower torque, than a 4-stroke will.

FUEL/OIL MIXTURES

Two-stroke engines made after the 1970s typically wear fuel caps noting that a gas/oil mix is required. Older 2-cycle motors might not so signify, but, if your engine has nowhere to put oil in its crankcase (not counting bar and chain oil reservoirs in chainsaws), it's most probably a 2-stroke model and will require oil to be mixed with the gasoline for proper lubrication.

Two-stroke oil additives are sold in both small containers that are premeasured for 1 gallon of gas, and in larger bottles that often have a measuring cup at their top for correctly dispensing the exact amount of needed oil into the gasoline. Ratios vary depending on the equipment and manufacturer. The most common ratio is 50:1, meaning that for every 50 parts of gasoline, one part of 2-stroke oil needs to be added. However, other mixtures such as 32:1 also exist, so be sure to consult your implement's owner's manual to verify the mixture you'll need. If in doubt and in a pinch, use the long-held standard of ½ pint oil to a gallon of gasoline (or 16:1 gas-to-oil ratio) to get you through in an emergency, as it will not harm your engine.

CLEANING A 2-STROKE ENGINE'S PORTS

Anyone who has experienced a miserable head cold knows how debilitating it feels to be all stuffed up. If your 2-stroke small engine could talk, after several years of use it'd probably blame symptoms of sluggishness and weakness on plugged-up exhaust ports. Carbon—especially from caked, unburned oil deposits—likes to clog exhaust ports and rob a 2-cycle mill of its spunky power and jack rabbit acceleration. Plus, excessive back-pressure from such clogs make a motor hard to start. While there's no cure for the common cold, the good news is . . . a 2-stroke can be remedied of its stuffy exhaust ports with little more than a sharpened stick and a few bursts of air.

Typically, access to these ports (often in the form of one to three holes drilled into the cylinder) can be gained by removing the muffler. Being careful not to let anything fall into the ports, unbolt/unscrew the muffler assembly and separate it from the cylinder. Doing so should provide you with a good view of the ports and their condition. It's not uncommon for a port in an engine that has been around a few seasons to be one-third to one-half obstructed by solid carbon.

A most important next step is to observe, via whatever view you might have through the ports, the piston and its rings. Once spotted, the piston needs to be in such a position as to prevent junk from getting into the cylinder or crankcase. Very slowly pull on the starter cord or (if that assembly has been taken off) rotate the flywheel manually. Do this until the skirt or near-bottom wall of the piston is covering the ports. At this point, you probably won't see the piston rings, as they'll be beyond the ports and towards the top of the cylinder. Then, with a sharpened dowel slightly smaller than the port(s), scrape away the carbon. If the stuff is really stubborn, a plastic scraper can be used. Screwdrivers or punches, gently guided, will work, too, but delicately so that the piston doesn't sustain any scratch that could compromise engine compression. When the solid carbon has been broken up, blow out the particles, ideally, with compressed air, being cautious not to get any of it in your eyes.

Now inspect the muffler. In older engines that were built during the do-it-yourself era, these units were essentially hollow boxes and typically able to be disassembled for cleaning. It's probable that yours is filled with sound deadening/spark arresting innards and has "no user serviceable parts;" a euphemism for: You need to buy a new one. Do so if your muffler looks clogged and ratty. Otherwise, give it a good soaking in solvent, lightly knock it with something that might help loosen the crud, let it thoroughly dry, and reattach it to the cylinder that now has nice clean exhaust ports. It isn't unusual to fire-up your "freshly serviced" 2-stroke and hear it speedily thank you for making it feel young again.

Exhaust port

Clean the debris and bits of old gasket before installing a replacement—take care not to let anything fall into the opening.

You'll need to remove the carburetor to get at the reed valve for inspection and replacement (if necessary).

CHECKING THE REED VALVE IN A 2-STROKE ENGINE

Pets love the freedom of those little spring-loaded doors installed in the base of a conventional one that allows them to simply push either in or out of the house at will. For small 2-stroke engines, however, such liberty aft of the carburetor would let ready fuel/air mix headed for combustion decide to make a hasty escape out of the front of the crankcase. On some larger 2-cycle engines, it's the job of a wafer-thin gatekeeper called a reed valve to keep the fuel mix on its proper pathway and prevent any loss in crankcase pressure, vital in the fuel induction and compression process.

Widely introduced in the mid-1930s on Evinrude outboard motors, reed valves are rather carefree unless something such as corrosion or an inadvertent poke with a screwdriver bends or breaks them. Not often found on newer small engines that power string trimmers or chainsaws, they are occasionally found on older or larger 2-stroke outdoor power equipment engines. Exact placement varies from engine to engine, but reed valves are usually either affixed to a plate directly to the outside of the crankcase (in back of the carburetor mounting flange), as part of a removable reed plate assembly "block" that fits into the front of the crankcase, or screwed to a reed plate attached to the inside front of the crankcase.

Often, reeds have a rough edge and a smooth edge, with the latter installed towards the reed plate. Carburetor cleaner solvent may be used to clean the reeds and related areas. Treat the reeds as delicate instruments being careful not to bend them. A feeler gauge (like the one for checking spark plug electrode gaps) should be used to measure clearance between the reed and its plate. While specifications differ, many 2-stroke engines work best with a reed clearance of no more than .015 inch. Rather than attempting to "re-bend" reeds that are out of tolerance, new ones should be installed.

CHAIN SAW MAINTENANCE

It's a good bet that a chain saw that either won't start or finally does, but runs and cuts poorly is a victim of caked-on oily sawdust and dirt, with at least a pinch of bad fuel to top off its troubles. After an hour or two of TLC in the form of de-gunking/cleaning the air intake filter, carburetor, and servicing the chain, most old saws will be buzzing away at the woodpile. The process often includes seeing if the saw's spark plug/ignition system yields spark, emptying/cleaning the fuel tank and fuel line from tank to "carb."

Because chainsaws typically employ a 2-stroke engine and a diaphragm carburetor (as opposed to one with a gravity/fuel activated float), they can literally be run upside down. This diaphragm is a paper-thin piece of synthetic rubber about 2 inches square that tends to get brittle or stiff (thus loosing its effectiveness) with age or exposure to harsh conditions. "Carburetor kits," a staple of small-engine shops' parts departments and something veteran saws hope to eventually enjoy, often contain replacement diaphragms and related gaskets, seals, needle valves, and specific installation directions for renewing carburetion.

Chain saws require diligent cleaning to run smoothly.

BAR/CHAIN OIL PUMP

The bar/chain oil system should also be on a chain saw owner's maintenance roster. It pumps lubrication from a tank to the area where the chain rotates in the groove in the chain bar. Note that this lube is not related to the oil required for the 2-stroke engine's gas/oil fuel mix. Like any other machinery, this bar/chain oiling system needs to be free of dirt and debris wanting to block or reduce oil flow. Of course, the more oil squirted onto the bar and chain, the more sawdust and debris will stick to them. That's why keeping the saw clean is so critical. Some pumps work automatically off a cam in the crankshaft, while others simply utilize a plunger pump a bit like the ones that dispense hand soap in a fancy bathroom. The older, manual type is sometimes fitted with a bleeder screw that allows adjustment for improved hydraulics. There's likely to be an air screw on automatic oilers, too. No matter the style, though, the pump output is in the vicinity of the bar mounting bolts and centrifugal clutch that drives the chain. That means access for cleaning requires removal of the drive clutch, a task that might present a bit of a conundrum because turning it (typically screwed onto the crankshaft via a left-hand thread) rotates the crankshaft along with the rest of the engine's main parts.

Removing the rewind starter to be able to get ahold of the flywheel will be necessary to keep the crankshaft still so that the clutch assembly may be removed.

Before proceeding, be sure that the groove on the circumference of the bar (in which the chain guides ride) is free of obstructions. Run the tip of a clean standard slot screwdriver in the groove and around the bar to check for blockage and/or "pinches" in this groove that would impede oil flow and increase chain friction. Clean and straighten accordingly. Look for/clean any passages running from the groove into the bar. Next, unscrew the pump assembly (often plastic) and gently pry it loose (along with the related oil line) from the saw with a flat screwdriver. Give the unit a thorough cleaning, then reverse the process to reinstall the pump.

BAR/CHAIN-GUIDES

Though a bar can become distorted from its saw being dropped or wedged into whatever it's cutting, the chain is more often the culprit when the old expression, like a hot knife through butter, is not fitting. The best way to inspect for "bar bend" is like one would do in order to determine if a 2 × 4 piece of lumber is straight; look at it across the top from engine to tip. While experienced machinists could probably perfectly straighten a bent bar, it is recommended that a bar with any twists or cracked, broken, or squished chain guide grooves needs replacing. The same suggestion applies to a bar with a tip gear that no longer spins freely. If the chain bar fits both ways, even with the manufacturer's logo reading upside down, it may be "flipped" so that the down side (where much of the chain pressure is exerted during cutting) can serve at the top, thus evenly sharing the wear over time.

CHAIN

Be aware of chain tension on your saw. Consult the owner's manual for proper tension specifications. Note that a chain running too loosely on its bar/chain guides will really wear into the bar, especially at the heel and just under the nose. The chain can be tightened by loosening the two nuts on the bar bolts (sticking through the engine crankcase), adjusting the bar forward, and retightening the nuts.

A $^5/_{32}$ round file, a flat file, and guide gauge are essentials when sharpening your saw's chain blade. Mounting the saw (via its bar) in a sturdy vise is one way to have good access for the procedure, though sharpening the unit while it's on a workbench or the ground is commonplace. In any position, however, be sure the chain can rotate freely. That accomplished, place the chain guide over the chain to gauge the desired angle (typically 30 or 35 degrees), remove it and use the round file to treat the front/cutting edge "tooth" of the chain at the determined angle. Remember that the sharpening action of the file should be in the "push" direction, not the "pull." Keep that file level, too. Use the gauge after treating each blade tooth to test whether it has been filed to the needed angle. Some chain saw files are fitted with the angle guides for ease of operation. Feel free to draw a colored marker line on each now newly-shiny tooth that you've filed so you can see what's been completed. Routinely, clean off the file so it doesn't fill with metal particles and oil. That flat file can be used to file down any chain guide that protrudes beyond the guide gauge. To file teeth running opposite of the ones you just sharpened, turn the saw around 180 degrees. No matter the side, treat every tooth with the same number of file runs, so the teeth won't end up being various sizes.

Chain tension is adjusted by loosening the bar bolts and moving the bar forward or backward.

Keeping your chain sharp requires a round file with a guide gauge.

STRING TRIMMER AND LEAF BLOWER MAINTENANCE

Maybe it's because those little gas trimmers and blowers can sometimes be bought new for not much more than the price of a couple of good shovels that they don't get the care they deserve. If doing nothing else, a trimmer owner should commit to keeping his/her unit clean with a rag. Special attention should be devoted to the engine air filter, which even if partially blocked with grass clippings, dust, leaves, etc., will cause the trimmer's (typically) 2-stroke engine to run poorly—if it even starts at all. An engine that sounds like it's running slower than normal and lacks power could be suffering from air intake obstruction. Foam filters can be washed in grease-cutting soapy water, but need to be completely dry before re-fitting to the carburetor. The exhaust/muffler should be free of obstructions, too.

If you've not run the trimmer in a while, check the fuel color. Particularly dark fuel—as opposed to a fresh translucent mix of gasoline and 2-stoke oil—is a bad sign and should be drained before a clean fill is facilitated. Good fuel flow is essential on a trimmer. If you see air bubbles in the fuel line, it could have water in it or the line could be cracked and allowing air. White or grayish-white exhaust smoke is a telltale of water or some kind of debris in the fuel/line.

At season's end, drain the fuel from the tank and then run the engine until the carburetor is empty. Introduce a few squirts of oil into the spark plug hole and pull the starter a couple of times to nicely distribute the lube on those moving parts about to take a long winter's nap.

And now a word about the other end of a trimmer: The part that is asked to do the dirty work. Check the cutting string (or blade system) and replace the string spool before it has a chance to run out or break apart on the job. Many string heads can be removed from their trimmers by pushing inward and turning the unit counterclockwise until it can be clicked/pulled off. Those without the patience to do things like thread a sewing needle or untangle fishing line understandably avoid winding bare plastic cutting string through trimmer head guides. They shouldn't feel bad about spending a few extra dollars on a complete replacement string cartridge (look for a heavier grade of line, if available for your model) that can be popped into place, as long as the trimmer head has been kept reasonably clean and tidy.

At season's end, always drain the gas from your string trimmer so you can start fresh next year.

Index